MODERN SERIES MADE EASY

PROCESS SAFETY MANAGEMENT IN OIL AND GAS UPSTREAM AND DOWNSTREAM

MODERN SERIES MADE EASY

PROCESS SAFETY MANAGEMENT IN OIL AND GAS UPSTREAM AND DOWNSTREAM

Authored by

ENG. GABRIEL GYAMFI

WINGS PUBLICATION INTERNATIONAL
WE GIVE WINGS TO YOUR DREAM!

Disclaimer

WINGS PUBLICATION INTERNATIONAL
WE GIVE WINGS TO YOUR DREAM!

Registered Office- 907-Sneh Nagar, Sapna Sangeeta Road,
Agrasen Square, Indore – 452001 (M.P.), India
Website: http://www.wingspublication.com
Email: mybook@wingspublication.com

First Published by WINGS PUBLICATION 2024

Title: **MODERN SERIES MADE EASY**
Price: AED 114 | $33

ISBN 978-93-6006-437-2

Special Dedication

This book is uniquely designed by Eng. Gabriel Gyamfi as a technical content to represent his commitment to serving humanity and it is dedicated specially to serve health and safety practitioners, learners, organizations, trainers, and oil and gas industrial professionals who will utilizes its 30 Elements intended to comprehensively foster the creation of more safer working environment in our modern communities through which limitless wealth creation and transfer of wealth can be actualized in human societies for the benefits of all.

Acknowledgement

My special gratitude goes to multiple individuals whose name may or may not be mentioned in this text, yet their immense contribution forms part of the pillars and foundation of which the creation of this book and its content has been made possible.

Designed by Gab-bys Creation, NO success story of an individual is ONE person's achievement.

My deepest gratitude goes to:

Mr. Sean Ferres (Group CEO, Ferres Enterprises, Author, Coach and Client Acquisition Consultant) Queensland, Australia

Dr. Kailash C Pinjani (Director, Author, Consultant, Success and Business Coach) He is also a Publisher, Founder and Chairman of Wings Publication International Enterprises Group

Mrs. Manika Singh (Author and Group CEO Wings Publication International) Headquarters and Global Subsidiaries

Dr. Deepak Parbat, An Author and Publishing Consultant (Wings Publication International) India, Pune

Mr. Abhas Mungal, Publishing Specialist, at Wings Publication International (Pune, India)

Dr. Ahmad Al Attas, An Author, Sustainable Leadership Consultant and Expert in Procurement, Logistics and Supply-chain Management, United Arab Emirates

Alex Hormozi, Author, Coach, Founder and Chairman of Acquisition.com and its Global Subsidiaries

Daniel Throssell, Author and CEO of Persuasive Page Pty Ltd, Australia

Mr. Naif Anwar Dhafari, Support Personnel, Integrated Transport Centre (ITC) Abu Dhabi, UAE

Mrs. Precious Yenduma Yakubu, my lovely wife Accra, Ghana

Mr. Rodel Cuntapay Passilan, Abu Dhabi, United Arab Emirates, UAE

Mrs. Grace Asiamah Business Woman Abu Dhabi, UAE & Accra Ghana

Mr. Anthonio Ahmed Adu-Gyamfi (Mr. AAA) my son Accra, Ghana

Mrs. John Sarfo Central London, United Kingdom of the Great Britain

Mr. Joseph Mensah Terku New York, Bronx, United States of America

Mrs. Comfort Oteng My Beautiful Mother, Nsuta-Kyebi, ASH.R Ghana

Engr. Rtd Mr. Yakubu Laari, My father-in-law Ghana Railway Quarters, #43 Affia-Kuma Secondi Takoradi (Western Region) Ghana

Mrs. Adiza Amadu, Business Woman, My mother-in-law Ghana Railway Quarters, #43 Affia-Kuma Secondi Takoradi (Western Region) Ghana

Miss Cecilia Ofori, Hair Stylish of Kumasi-Atimatim, Ashanti Region of Ghana

Mr. Lameck Manu, Business Manager and a Cyber Security Specialist, Abu Dhabi, United Arab Emirates

Mrs. Nicoline Kalle Eleh, CEO Blessing Properties Abu Dhabi, United Arab Emirates (UAE)

Mr. Hussein Mutebi, Health and Safety Specialist, Abu Dhabi, UAE

Mr. Joseph Yugtii, at Kwame Nkrumah University of Science and Technology (KNUST) Kumasi-Ghana

Miss Adomako Mensah Abigail, of Agogo Presbyterian Wemen's College of Education, Ghana

Preface

Well, Well, Well

Dear Gentle Reader,

My name is preface, kindly permit me to gist you about what this author is saying concerning the most unique way of using well-structured designs in the **upstream** and **downstream** management of hydrocarbons production process under this **process safety management in oil and gas book meant to serve.**

Well, in the vast and intricate world of oil and gas, where the stakes are as high as the rewards, the concept of Process Safety Management (PSM) stands as a sentinel, guarding against the unseen perils that lurk within both upstream and downstream production processes. This preface embarks on a journey through the entanglement of safety protocols, risk assessments, and preventive measures that form the backbone of this PSM book to serving oil and gas industry globally.

Well, from the moment crude oil is extracted from the depths of the earth in upstream operations, to its transformation into refined products in downstream facilities, the potential for hazardous incidents is ever-present. The upstream sector, with its drilling rigs and offshore platforms, faces unique challenges such as blowouts, well control issues, and equipment failures. Here, this PSM book is not just a regulatory requirement but a critical practice that ensures the integrity of operations and the safety of personnel.

Well, as the journey continues downstream, where crude oil is processed into valuable products like gasoline, diesel, and other petrochemicals, the complexity of operations increases. Refineries and chemical plants are rife with high-pressure systems, flammable materials, and intricate processes that demand meticulous attention to safety. PSM in this realm and the content of book involves rigorous hazard identification, risk management, and the implementation of robust safety barriers to prevent catastrophic events that are preventable in the oil and gas sectors within the global energies production.

This preface aims to enlighten your curiosity about the multifaceted world of Process Safety Management in the oil and gas industry. It invites you to explore how a disciplined framework, grounded in engineering principles and operational excellence crafted by this author, can mitigate risks and safeguard both human lives, company assets and the environment. Join us as we delve into the essential practices that uphold the safety and integrity of one of the world's most critical industries.

Contents

SECTION I ELEMENT ONE – FIVE

Fundamentals of Health and Safety in Oil and Gas Industry

ELEMENT ONE: FUNDAMENTAL DEFINITIONS OF HEALTH AND SAFETY

ELEMENT TWO: SAFETY IN THE OIL AND GAS INDUSTRY

ELEMENT THREE: THE HEALTH, SAFETY AND ENVIRONMENT LAW

ELEMENT FOUR: THE LENGTH OF THE LAW

ELEMENT FIVE: METHODS OF COMPLIANCE

SECTION II ELEMENT SIX – TEN

Leadership Context in Process Safety Management (PSM)

ELEMENT SIX: MANAGING HSE SYSTEMS IN OIL AND GAS

ELEMENT SEVEN: OBLIGATIONS IN PROCESS INDUSTRY (PI)

ELEMENT EIGHT: MANAGEMENT OF CHANGE (MOC)

ELEMENT NINE: PROCESS SAFETY MANAGEMENT (PSM)

ELEMENT TEN : MANAGING CONTRACTORS

SECTION III ELEMENT ELEVEN – FIFTEEN

Plant Management and Process Industry General Operations

ELEMENT ELEVEN: PLANT OPERATIONS, INSPECTION AND MAINTENANCE

ELEMENT TWELVE: INSTALLATIONS START UP AND SHUT DOWN

ELEMENT THIRTEEN: PERMIT-TO-WORK PROCEDURES AND COORDINATIONS

ELEMENT FOURTEEN: ESSENTIALS OF SAFE SHIFT HANDOVER

ELEMENT FIFTEEN: FLOATING PRODUCTION STORAGE AND OFFLOADING (FPSO)

SECTION IV ELEMENT SIXTEEN – TWENTY

Safe Systems of Work and Risk Management Methods in Oil and Gas

ELEMENT SIXTEEN: SAFE SYSTEMS OF WORK (SSOW)

ELEMENT SEVENTEEN: HAZARD INHERENT IN OIL AND GAS

ELEMENT EIGHTEEN: SOURCES OF PSM INFORMATION

ELEMENT NINETEEN: RISK MANAGEMENT METHODOLOGIES

ELEMENT TWENTY: EVIDENCE DOCUMENTATION

SECTION V ELEMENT TWENTY-ONE – TWENTY-FIVE

Balancing Capacity, Safety and Efficiency in Hydrocarbon Storage

ELEMENT TWENTY-ONE: FIRE & EXPLOSION HAZARDS PROTECTION AND RISK MANAGEMENT

ELEMENT TWENTY-TWO: SAFE OPERATIONS OF FURNACE, BOILER AND STEAM MANAGEMENT

ELEMENT TWENTY-THREE: HYDROCARBONS SAFE STORAGE MANAGEMENT

ELEMENT TWENTY-FOUR: UPSTREAM, DOWNSTREAM AND MIDSTREAM ACTIVITIES

ELEMENT TWENTY-FIVE: EMERGENCY RESPONSE PROCEDURES

SECTION VI ELEMENT TWENTY-SIX – THIRTY

The Case Study, Environmental Impacts, and Tasks Assessment

ELMENT TWENTY-SIX: THE CASE STUDY

ELEMENT TWENTY-SEVEN: THE ENVIRONMENTAL IMPACTS

ELEMENT TWENTY-EIGHT: CAUSES AND LESSONS LEARNED FROM INCIDENT

ELEMENT TWENTY-NINE: TASKS ASSESSMENT (Multiple Choice Questions)

ELEMENT THIRTY: TASKS ASSESSMENT (Theoretical Questions)

Introduction

In the field of occupational safety and health, process safety management abbreviated as PSM is a unique procedure consisting a number of mechanisms designed to provide protection to the people in a society such as industrial workers, the environment and the properties of organization that work with or produces materials termed as a Highly Hazardous Substances.

The functioning of PSM in this modern series made easy edition has been designed to suit the needs of learners and safety professionals imed to match them with the industrial dynamics, technological advancement, local and international best practices and compliance with the regulatory bodies.

Due to the high level of hazards and risk associated with the operations in the process industry, this book contains six (6) different sections and under them lies all the elements which has detailed information of process safety management systems (PSMS) required for achieving industrial safety, employee's safety and health, environmental protection, and the safety of our entire communities where the oil and gas installations has been established and operates.

The book focuses on petrochemical industries especially, petroleum production companies who deals with hydrocarbons and their inherent hazards alongside the risks they posed to those exposed on daily basis in order to achieve safety for all.

The creation of this Modern Series Made Easy: Process Safety Management version is strictly based on international standards of best practices and thorough research on the health and safety needs of the global business entities in the energy sector that produces oil and gas.

Among other factors, identifying human errors as the culprit behind most of these catastrophic accidents, preventable fire and explosions, methane gas releases and others that has claimed thousands of human lives over the years, causing permanent disability, incurable diseases and sickness, damaged to company's assets and reputations have their remedies captured.

Other negative impact on the environment and the effects on other non-human living organisms in different ecosystems such as wild animals in the forest, aquatic lives in ocean and seas which always ended up in a very devastated consequence have their mitigations requirements provided.

The authorship of this book has considered all other factors that could result in disaster and has understood, most of these incidents are preventable with the Right Continuous Development in teaching and training materials suitable to create competent workers for the match of process industrial needs worldwide.

In view of this gave birth to the creation of this book, Modern Series Made Easy: Process Safety Management (PSM) in oil and gas to empower and equipped professional individuals and industrial entities operating in the hub of global energies, typically of hydrocarbons production processes to uplift the safety needs to the suitability standards of present times.

This author, Eng. Gabriel Gyamfi believes that, prioritizing Occupational Safety and Health, Advancing Environmental Protection and integrate the HSE into the corporate core functionalities at all levels in the process industries across multiple sectors in the global economic landscape would not only enhance the PSM, rather it would go a long way to increase production efficiencies in terms of quality and quantity at the same time, thereby increases Return On Investment (ROI) in the long-run.

Safeguarding human lives, workers and members of the public alike, preventing property damage of businesses while ensuring adherence to all compliance requirements at both national and international levels remains another major key to achieving a safer working environment in the process industry for all stakeholders and content of this book outlines how that could be achieved.

SECTION I

Fundamentals Of Health And Safety In Oil And Gas Industry

Upon completion of this section, participants would gain knowledge of the fundamental definitions of health, safety, welfare, accident, near miss, dangerous occurrence, as well as safety in the oil and gas industry.

The section also includes the driven force behind health and safety known as the "LAW" and the methodologies use to ensure compliance with the HSE regulations.

Take-Away benefits of this part consist of the following elements:

- **Fundamental Definitions of Health and Safety**
- **Safety in the Oil and Gas Industry**
- **The Health, Safety and Environment Law**
- **The Length of the Law**
- **Methods of Compliance**

ELEMENT ONE

FUNDAMENTAL DEFINITIONS OF HEALTH AND SAFETY

HEALTH

This refers to the act of protecting the bodies and minds of people (workers) from illness caused by workplace hazards, processes, or procedures.

The basic view of health in this case is to safeguard employees and other stakeholders from both physical and mental illness arising from any operational activities of an organization, typically, in the oil and gas industry.

SAFETY

Safety is the act of protecting individuals (employees) from physical injury. Here, the word safety can be extended to all the necessary measures in the form of policies and procedures used to identify workplace hazards and their risk associates to determine who can be harmed and how, and the needed elimination or mitigation mechanisms, using the hierarchy of controls or Control Hierarchy.

WELFARE

Welfare in general defines the good practices which involves the provision of basic facilities needed for employees' comfort in their various workplaces.

Examples of welfare include portable drinking water, availability of First aid and First aider(s), suitable washrooms (toilet and urine), designated places with warning or caution signs like Exit lights, electrical hazards, biomedical hazards, radioactive hazards signals and among others.

OCCUPATIONAL OR WORK-RELATED ILL-HEALTH

The occupational or work-related injury or sicknesses describe any form of illness that an employee may suffer such as physical injuries, mental disorders and other harmful effects resultant from workplace activities. For example, prolong inhalation of poisonous substances causing lungs to damage (lung cancer) or heart failure to workers.

ENVIRONMENTAL PROTECTION

In the simplest form, environmental protection involves any type of arrangement put in place with the sole aim of protecting the environment about Land, Water/Water-bodies, and Air (atmosphere).

In some cases, the good practices of these arrangements by some of the process industrial organizations can be extended beyond protecting the environment and factor into consideration the purpose of keeping safer environment.

This has to do with the company's ability to expand its environmental management and sustainability programs to cover the protection of all living organisms, namely human beings, animals, agricultural live forms and climate conditions.

For example, the environmental management system used by Abu Dhabi National Oil Company popularly known as (ADNOC) one of the global leading companies in energy production in the oil and gas industry has the coverage of land, water, and air protection, as well as the protection of the people, and other living things residing within the environment in which the company operates.

The ADNOC Group have also included climate protection to control emission and pollutants that contribute to the causes of global warming when release into the atmosphere in its environmental management and sustainability policies.

ACCIDENT

An accident refers to the occurrence of an unplanned event that has caused an injury to people, ill-health, permanent disability, property damage, and in worst-case scenario death.

For instance, in the year 1984/85 the release of poisonous gas in Bhopal India was a chaotic accident that claimed human lives, inflicted chronic sickness, leaving other health defects on people and a detrimental pollution to the environment's non-human habitats had been one of the saddest accidents never to be forgotten in the process industry (PI).

NEAR MISS

Near misses are similar to an accident but without any casualty. A near miss is an unplanned event that has occurred without causing any harm or injury but has the potential to cause harm or ill-health when unguarded, therefore needs to be investigated to provide appropriate risk control measures as a precaution to prevent recurring or similar related incident.

Conducting such investigations into incidents be it an accident or a near miss are necessary and must not be limited to the finding of both immediate and root causes, this way the findings can be used to risk-assessed the scenario for possible strengthening of weaknesses in existing control measures, thereby improving the overall workplace safety systems to the benefits all stakeholders.

DANGEROUS OCCURRENCES

These often relates to the happening of a specified event that needs to be reported to the relevant authorities concerned and they are usually guided by the requirements of the statute laws in a given geographical area or a country.

Occurrences such as earthquakes, volcanoes, tsunamis, mass fire outbreaks, Ebola, Corona virus, bomb threats, Flood and so on are classified under this condition.

The magnitude of dangerous occurrence often threatens the public health and safety of the society thereby needs to be addressed by competent authorities instituted by national or state regulations.

Any good practise of Occupational Safety and Health must be designed to include the provision channels to report such incidents/ accidents to designated authorized agency of the government when occurred without any hesitation.

ACCIDENT RATIO

Accident ratio depicts the existing relationship between the numbers of accidents that have occurred with varying degrees of outcomes.

Some multiple analysis indicates that, this relationship forms a triangle, showing the most serious outcomes at the top representing the highest, followed by those with a higher proportionality, and others classified as less serious formed the base.

The illustration below is an example of accident ratio triangle depicting the relationships numbers from the top-to-bottom respectively.

Incidents/ Accident Pyramid

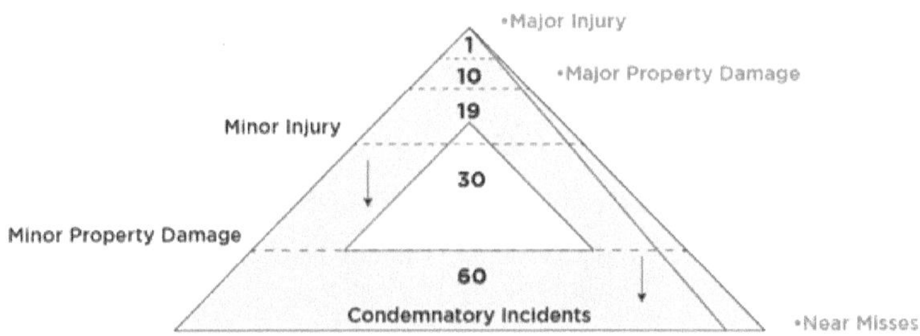

The illustration above shows different kinds of incidents with varying degrees of outcomes.

- Major Injury Accident,
- Minor Injury Accident,
- Major Property Damage Incident,
- Minor Property Damage Incident
- Near Misses Categories

The arrows pointing downwards, Indicate a decline in impact of those occurrences and must be reduced to the acceptable level also known as The Residual Risk.

OCCUPATIONAL HEALTH AND SAFETY REPRESENTATIVE

This refers to an individual or personnel, with professional qualifications and a qualified technical ability to anticipate, recognize, evaluate and control hazards that have a connection to work.

Selecting personnel to occupy such position should be determine by a safety term known as KATE. The term KATE is an abbreviation word which implies **K**nowledge, **A**bility, **T**raining and **E**xperiences gained by an individual.

Among other things, personnel must possess the above-mentioned criteria requirements as fundamental needs on the subject matter 'the JOB'.

This is because in the field of Occupational Health and Safety, anything less than perfect could be a detrimental causing harm or injury to the people, damage to the company's asset, damage to public and private properties and in serious cases, permanent disability or death.

SAFETY SUPERVISOR

The position of a safety supervisor in this context describes an individual tasked with the workplace responsibility for a supervisory role and has the authority of supervision and direction of the work team.

Such individual must possess the required competence on the field in addition to leadership qualities, efficient observation skills, and accurate judgement on risk and hazard-based situations in connection to the work.

A competent safety supervisor should also possess the ability to generate options by providing guidance to mitigate work-related conditions that are unsafe to workers, members of the public, and properties while ensuring the safe keeping of the environment at all times.

ELEMENT TWO

SAFETY IN THE OIL AND GAS INDUSTRY

The general thought of safety at various working environment can be linked directly to the natural way of an individual thinking about the safety of people or employees who can be affected in many ways at their place of work due to work-related risk of physical injury, mental disorder, or ill-health.

Several kinds of personal accidents can be categorized as common and simple, which could be considered reasonably foreseeable such that, measures to control them can be well-established to grant straightforward access and implementation at all times to eliminate or mitigate the hazards associated with them.

Typical examples include fire preventive measures namely, (fire extinguishers, smoke and heat detectors, and water sprinkler system), machine guarding, equipment and machinery maintenance and inspection, managing workplace slip and trips, the use of Personal Protective Equipment (PPE), and so on.

When compared to the standards, the term **"process safety"** is about the safety in high-hazard process industries which are more complicated due to the high level of risk factors such that industrial organizations like chemical manufacturing units, oil and gas companies, refineries and among others are classified as high-risk ventures.

While many working-class in other industries across different sectors worldwide are being exposed to personal accidents at their workplaces, there is a variation that workers in oil and gas, and chemical production companies have the potential to encounter a major incident.

This is because employees in such industries deal with a larger volume of chemicals that are highly dangerous to health and safety in their production processes at all levels as such safety in oil and gas industry should be managed with high level of precision and accuracy in terms of supervision, controlling and monitoring.

Inadequate monitoring and control of hazards in process industrial units, especially chemical plant and oil and gas can easily lead to a spectacularly wrong, causing major fire outbreaks or explosion and releases of poisonous substances.

Though terrible incidents of such nature are very infrequent adversity and sometimes, it can be difficult to predict as in how, what, where and when it will happen.

These prediction difficulties exist in those installation because of their complex nature, and the involvement of multiple factors that causes them, including not taking into account or neglecting the things that seem small could be the actual culprit responsible for a major catastrophic incident.

In line with process safety, the emphasis on the **"prevention of major disaster"** can be seen to have a historical issue in the industry where company's operational processes are classified as highly hazardous.

Process safety requires both a robust safety management systems and complex technical controls, in addition to competent personnel whose performances are balanced against industrial standards, compliance and high level of precision and accuracy to attain safety for all.

Given this, the general knowledge requirements should be based on an individual with expertise in technical engineering and a good deal of management skills to achieve desirable results and the leadership in oil and gas organizations are expected to make it their utmost importance to prioritise **process safety** at all times.

Standard and controls indicate that, incidents should be unusual and may occur beyond the level of operator's experience, **personal-safety** and **process-safety** are analyzed separately since they do not mingle together under the directions of process safety management (PSM).

The main aim focuses on the prevention of the high-risk large-scale destructive events that may occur such that, the likelihood of this catastrophic adversity being highly rare, its occurrence could result in very devastating consequences.

According to the Centre for Chemical Process Safety of the United States of America, the **American Institute of Chemical Engineers Process Safety** is the combination of engineering and management skills that focused on preventing catastrophic accidents and near misses.

Their area of interest includes, structural collapses, explosions, fires, and toxic releases resulted from the loss of containment of energy or dangerous substances like chemicals, petroleum products, and hydrocarbons gases.

ELEMENT THREE

THE HEALTH, SAFETY AND ENVIRONMENT LAW

Introduction

The purpose of examining safety and health in the construction industry which has spread across multiple sectors in recent times could be considered unrealistic and inadequate without the support of the laws behind them and it is the law that ensure rules are enforced, requirements are met against standards and compliance achieved.

The functions of the law when analyzed thoroughly, with all the legal implications that are likely to affect some of the participants, then engineers should be made known of the basic nature of their duties and responsibilities to the law.

Typically, on liabilities in both criminal and civil cases that are relative to the two main sub-divisions of the law which applies to health and safety-related issues and in most countries, the arm of government establishes department tasked with the responsibility of the Judicial administration.

This involves the Justice System of the community, the Court Structures ranging from the highest level to the lowest, and other relevant institutions, for instance, prison services (Punishment and Correctional Facilities).

THE CRIMINAL LAW

This usually refers to the various rules and regulations that are constituted often by the Government, or the State, or any Govern Body represented as the Competent Authority.

They are capable of enforcing the society's participants to be law-abiding, mostly enacted by the Parliament through the Acts of Parliament and the imposition of these rules or Acts on the people residing on an area or in country has the sole aim, which is to protect the people.

Criminal law describes all of the well-established constitutions, statutory requirements, and legislation of different levels, enacted by the highest decision-making body, and can also be formulated under the Eexecutive, Legislature, or the Judiciary in broader context of running a State.

It serves the purpose of regulating the behaviour of society. Such laws are enforced by multiple Government Mechanisms that serve different functions to ensure human lives, properties, and information are preserved and protected.

These government agencies reserve the right to conduct investigations, including the prosecution of individuals and organizations for criminal charges due to contravening the requirements of the criminal laws.

In this case, the offense committed is referred to as "Crime" (which is the offense against the State), while the word "Criminal" implies a common name given to an individual or a group, or corporate entity that committed the Act.

Even though criminal laws and punishments may vary between countries, the common norms indicate that only these government agencies can determine whether or not if a crime has been committed, and the culprit individual or a company should be prosecuted.

Likewise, the required punishment is prescribed by the law, with the exception very rare cases because, if a person or company breaks criminal law, they become liable for criminal charges that would be determined by a competent court of law under a specified jurisdiction.

This is done through the prosecution processes, where the offender an individual or business entity will be found guilty for the crime committed and being guilty can only be determined according to the law requirements, including the required punishment.

For example, imprisonment sentence (that is where the offender serves a jail term in punishment and correctional facility).

A fine to be paid as a monetary compensation pronounced by the Criminal Court is the sentence rendered as an alternative to the jail term, and the verdict could also include both fines and jail term to be served.

Being an employer or employee, owing to the possibility of the loss of freedom determined by a competent authority requires a high level of proof by the Criminal Court during proceedings that led to the judgment passed.

The Criminal Court decisions are based on the facts such as Elements of Crime, Elements of Guilt, and must be supported by proof of evidence provided before ruling.

Establishing the guilt of a crime is directly related to the court sentence bringing the verdict to the **very near certainty as possible**, known as **"beyond reasonable doubt"** in legal terminology.

In generic terms, the court structures and legal systems are highly identical such that, criminal courts in most societies were established with the **Prime Objective** of allocating punishments, the fear of punishment primarily serves as the deterrence and discouragement to the members of the public to avoid breaking criminal laws.

Despite such prime objective, the Criminal Court can award huge sums of money as compensation to the worker **(victim)** or injured party who might have suffered a great deal of loss due to a work-related accident.

Detrimental losses like Permanent Disability, Body Part Amputation, for example, (leg, hand, etc.) serious degree fire burns or the unbearable scenario case like death could be treated as crime.

THE CIVIL LAW

This concern with all the rules and regulations established by the competent authority, usually the government and its regulatory bodies purposely to address disputes that may arise between individuals, groups, and business entities in the society.

Under these laws, a person may sue another or an employee suing a company or employer or a firm suing another corporation, requiring a formal settlement of wrong or tort civilly committed against them without any involvement of criminal charges.

The intended purpose of this law is to ensure that justice, fairness, and equity has been administered among the participants of the community who are involved in civil cases.

An individual or organization who filed the lawsuit in the court seeking a fair judgment is called the "Plaintiff, Claimant, or Complainant" while the other person or a company being sued is known as the "defendant".

The civil laws, in this case, give the right to Civil Courts to administer all the justice requirements with little or no regard to the guilt or non-guilt, instead, the court concerns itself with liability and the exact extent of that liability in question.

However, the determinants of the civil law at this level during court proceedings for judgment are based on proof be it directly or indirectly related to a specified **"balance of profitability"**

This may be considered as a lower level of certainty when compared to the level of certainty required as a proof by the Criminal Court before ruling on a case presented.

In the event where a defendant is found liable, the court would then order such individual or business entity to pay compensation often in monetary terms to the plaintiff, and the defendant may be referred to as "tortfeasor"

Tortfeasor is the legal terminology that best described an individual or a corporate entity found to have committed a civil offense that injured another person or caused property damage or information etc.

Sometimes, this compensation payment may also include financial losses incurred by the claimant as determined by the court upon giving the verdict, and there is a common uniformity in civil laws relating to the term **"balance of profitability"** which can be summarized as the Balance of Profitability should be equated to:

- the lower the balance of profitability, the lesser the level of monetary compensation awarded, and
- the higher the balance of profitability, the greater the level of monetary compensation awarded.

In rare cases presided over by the civil court, where the balance of profitability is highly competitive, just as over 50% among the parties involved, the plaintiff may "win" the case alright, but suffers financial losses because awarding costs incurred may not be included in the ruling, thereby bringing the compensation to a low level.

In most cases involving health and safety, Civil Court settle such disputes often follow workplace accidents or ill-health caused by negligence act or a breach of specific statutory regulation.

The larger part of these cases are resolved **"out of court jurisdiction"**. The most common form of such cases is the actions between individuals, where the defendant is a worker who was performing his or her duty assigned by employment contract at the occurrence of the alleged incident.

Normally, the defence of such action is transferred to his or her employer. Under this condition, the civil action would then become a case between the plaintiff and the employer and was considered by civil laws as vicarious liability.

VICARIOUS LIABILITY

In a society where the proper function of legal systems categorically separates Criminal Law from Civil law and the structure of court allows the independent functioning of the law of crime and the law of tort, then the civil law administration should have a provision that covers vicarious liability.

Vicarious liability defines a practicable doctrine through which an individual often (employer) is held liable for the tort committed by another person usually (an employee) against a third party.

The principles of vicarious liability make it possible for the employer to be liable for the torts of his or her workers who were acting in the line of duty that led to the incident and during court proceedings involving vicarious liability, the following legal terms may be used to represent the individuals involved.

1. First Party (the employer),

2. Second party (the employee),

3. Third party is the injured / affected individual

The term **"third party"** in this context can be represented by different individuals such as an employee's co-worker in the same organization, an employee of another company working on the site, or a member of the public.

Regardless of who may suffer injury due to the negligent act committed by a worker that causes an accident at the time he or she was performing duty function required by the employment, the doctrine of vicarious liability, command employers to be liable.

Vicarious liability gives no regard to whether or not if the worker was acting on the express permission or authority given by the employer, any ssuccessful claims made by the injured person due to work-related accidents were made possible on the ground of Vicarious Liability.

For example, the driver of a commercial vehicle involved in an accident is not held liable for damages caused by the vehicle or caused to the vehicle or both, rather the owner of the car who receives daily sales is made liable by the terms of **vicarious liability,** which in most cases losses incurred are settled by insurance coverage, where applicable.

ELEMENT FOUR

THE LENGTH OF THE LAW

The laws associating with the Occupational Safety and Health can be extended to most of its pioneers dated in **1919** and **1974.**

Acknowledging the full length of the law and other requirements of Health and Safety, and the general welfare of employees at the various workplaces would require broader references from several bodies existing at both national and international levels.

The success of Health, Safety and the general well-being of workers have the following bodies that are well structured, with the core aim of providing a safer working environment for all.

- The International Labour Organization (ILO), Parental Organization the United Nations, the (UN) inception **1919**
- The Occupational Safety and Health Administration (OSHA), the United States Federal Government | Parent Organization United States, Department of Labour **(1922), 1971** as Bureau of Labour Standards
- The British Health and Safety Executive (HSE), UK. | The Health and Safety at Work, etc. Act 1974, (HSE Act). The Parent Organization, the Parliament of the United Kingdom/Great Britain (Royal Assent).
- European Agency for Safety and Health at Work, (EU-OSHA) Parent Organization, European Parliament / European Commission /

European Union-EU **(1994)**

The establishment of the UK's Health and Safety at Work etc. Act, in 1974 (HSE Act) was primarily to cover all people at the workplace including self-employed, employers, employees, and by extension to offer health and safety protection to the public at large who may be affected by work-related activities.

The coverage also includes other working parties at the site who are not direct employees of the organization such as contractors, vendors, suppliers, etc.

The Act of Health and Safety at Work instituted by the Parliament of Great Britain in **1974** serve the territories, of England, Wales, Scotland, Northern Ireland, and Offshore by extension.

To discern successful implementation, the HSW Act came into effect in four different parts and two bodies were tasked with the supervision responsibilities, the Health and Safety Commission and the emergence of Health and Safety Executive the UK's (HSE).

The Health and Safety at work etc. Act 1974

PART 1

Under this part, the creation of health and safety at work etc. Act of 1974 is explicitly concerned with safety in the construction industry in the region or geographical area it covers. Here, the purpose of this first (1st) part can explain below as follows:

- Guarantee the health, safety as well as general welfare of the people at work
- Ensure to protect other people who are not physically involved at work against risks to their health or safety occurring from the activities of the individuals at work.

- Implement adequate Control for the keeping and use of an explosive or highly flammable or any other Dangerous Substances and generally ensure to prevent people from acquiring and using such substances unlawfully.
- Prevent the release of all Noxious or Offensive Substances into the atmosphere, including keeping them away from the premises prescribed by the regulations.

The continuation of the HSW Act part 1, from the second (2nd) section through to Ninth (9th) categorically States, the General Duties assigned to both Employers and Employees.

The regulations prescribed in section two (2) that every single employer must ensure the health, safety, and welfare at the workplace of all his or her workers are adequate especially in the following areas:

- Provide arrangement and maintenance of Plant and Systems of work that are safe and also without risk to their health.
- Making provisions required for ensuring safety and eliminating the risk to health that could result from the Use, Handling, Storage, and the Transport of Articles and Substances.
- Providing all the essential details regarding Information, Instruction Training, and Supervision to employees for safety awareness.
- The preservation or repairs of workplaces under his or her control as an employer is in a safe condition and without health risk, and must also include the provision and maintenance of a working environment that serves as a means of Access to and Egress from the said workplaces are safe and without health risk.
- Employers are required to continuously provide and maintain their working environments to be in safer conditions and without risk to health that is generally considered adequate based on the facilities made available to facilitate their employee's welfare.

- Apart from the above-written regulations prescribed, the end part of section 2, outlined the size of an organization that must adhere to the instructions stated by the Health and Safety at Work, etc. Act of 1974.

Under this section, all employers with a headcount of workers ranging from five (5) or more must comply to put together a (Written Statement) of their general policy for the health and safety at work suitable for the business entity they operate, that would safeguard their employees and others.

- Among other things, the written statement of the company must include but not be limited to the organization's laid down process or procedures for carrying out that intended policy of the statement, as well as a certain period or date set as necessary to revise the policy to ensure it matches the current status of the entity.
- The end part of section 2 of the Health and Safety at Work etc. Act of 1974, tasked employers to establish an adequate channel to communicate the content of the policy statement to all the employees.

The third (3rd) section requirement of the HSW Act entails that every employer must conduct their undertakings in a manner to ensure that, other individuals outside of his or her employment are not exposed to the risk of their health and safety by the operational activities of the said undertaking.

Alternatively, any individual (natural human being) who own a firm and acted in the capacity of self-employed must as well conduct his or her undertaking in a way that they and other people in the society are not exposed to their health and safety and this particularly contains the provision aimed at protecting all the people in the community.

This regulation placed a burden on business owners to ensure the entire population within the geographical region where every activity of their undertakings must be safeguarded to protect the public health and safety rather than causing risk exposure to them by the work activities.

The fourth (4th) section consists of regulation towards the imposition of duties on those who by any reason have control over premises.

Their obligations include the means of accessing, Machinery, Plant or Substances under their control that are being used by individuals who are their employees at the work environment.

Here, the control of premises has the duty of ensuring all the required measures are taken to provide all users with safety and without health risks.

The Fifth (5th) section in the regulation requires all those who control premises to exercise the best application means for preventing any emission of noxious or offensive substances into the atmosphere.

This requirement placed duties on the controller of premises to ensure the rendering of harmless, and inoffensive activity on any poisonous substances and do not allow escape into the human environment to guarantee safer atmospheric conditions.

The sixth (6th) section of the Health and Safety at Work etc. Act 1974, categorically places duties on persons under a certain bracket whose activities are inter-related.

These are designers, manufacturers, and importers or suppliers of various articles for use at work. Individuals under such bracket can be classified as the **"Supply Chain".**

The requirements of section six (6) arc as follows:

Anyone who design, manufacture, import, or supply articles for use at work:

I. Must ensure that their article has been designed and constructed in a manner that it is safe and without risk to health when properly used according to the intended purpose to which it was manufactured.

II. Must provide arrangements for any necessary testing and examination of their articles; and that, designers and manufacturers are required to make provisions for all the necessary research needed with a view to the discovery, to eliminate or minimization of any risk to health and safety to which the design of their product may cause.

III. Must include steps taking needed to ensure adequate information will be made available about the use for which their article has been designed and tested.

IV. Must ensure that, if they are erecting or installing any article, in premises to be used by individuals at work, that article is installed or erected safely and not a health risk when properly used as required.

The seventh (7th) section of the regulations goes directly to employees while at work. Here are the requirements placed on the shoulders of workers:

1. They must be reasonable in taking care of their health and safety at work at all times, including those who may be affected by the actions of omissions or mistakes arising from discharging duty and responsibility assigned at the workplace.

2. They must co-operate with their employers and other relevant individuals as necessary enabling them to comply with all statutory requirements placed on them.

The eighth (8th) section of the HSW allocate some level of duty

to all populace, under varying degrees of categories such as those at work, members of the society, and including children; Such

That they must not misuse anything that has been provided on the intent of health, safety, and general welfare and not intentionally interferes with an intended purpose to which article has been designed and constructed for use by the manufacturer.

The ninth (9th) section of the regulations is meant to ensure that, employers are prohibited from changing their employees regarding anything done, or needed to be done purposely for providing health and safety and to comply with particular statutory requirements.

The remaining forty five (45) sections of PART ONE (1) of the "Health and Safety at Work etc. Act 1974" are the provisions stated as powers and duties of the Health and Safety Commission, **Health and Safety Executive, and their enforcement, offenses,** and all other miscellaneous related issues.

By the Royal Assent and the Parliament of the Great Britain conformance, the successful implementation of the regulations in the Health and Safety at Work, etc. Act 1974, abbreviated as (HSW), and embodied the provisions of certain codes of practice.

This code of practices falls under sections 16 and 17 of the HSW etc. Act, which indicated that the Health and Safety Commission is responsible for the issuance of, and approved codes of practice needed for safer work processes at the workplace.

Additionally, the Health and Safety Commission reserve the right to approve codes of practice whether prepared by itself, or by other bodies of competence recognized by the law.

For example, the British Standards Institutions can prepare a code of practices with the sole aim of providing practical guidance based on the requirements of some part of the HSW. etc, Act typically from the 2nd to 7th sections and not limited to any other health and safety regulations or statutory requirements.

Non-compliance or failure to observe the guidance provided as an approved code of practice is not in itself unlawful.

Instead, if an individual or entity is being prosecuted for any breach of requirements stated by the Health and Safety at Work, etc. Act 1974, especially from the second 2nd to the seventh 7th sections, or other related legislation, then a certain approved code of practice relevant to that breach can therefore be used as evidence (relevant code is admissible as evidence).

It would be the defendant's duty to show whether or not if that relevant code of practice was complied with, or what took place was as good as or even better than the stated provisions in the code.

The Justification of compliance or noncompliance, in this case, is relative to the prescribed method of compliance.

ELEMENT FIVE

METHODS OF COMPLIANCE WITH THE LAW

These are methods to ensure that approved codes of practice provided by the **Health and Safety Commission** are adhered to by those tasked with duty and responsibilities stated by the Health and Safety at Work etc. Act 1974.

These compliance methods describe different ways that are used by the **"Field Inspectors"** deployed by the Health and Safety Commission to secure compliance with legal requirements in the country.

Such method requires frequent inspections of workplaces that need to be made by designated field inspectors trained with the norms of health, safety, and welfare regulations and breaches.

Under this condition, most of the compliance methodologies involve unannounced or uninvited routine inspection of sites to ascertain whether work activities being carried out are in conformance with the law requirements.

Such visits to workplaces by inspectors without giving any prior notice to the contractors, or site managers allow the occurrence of the breach to Health and Safety regulations to be identified.

The prescribed action would be taken against the offender whether or not it is an individual, a group or business entity. Ensuring that compliance is met, the field inspectors follow a series of procedures shown below:

• **Advisory and Persuasion method:** The field inspector during his visit to the site or workplace without warning notice to the contractor or site manager may identify some level of contraventions, for instance, the Construction Regulations.

Some of these cases occur, because specific requirements may be unknown to the contractor or site manager, or how they may be complied with, under such conditions, the inspector will provide perfect advice on procedures to which the obligation would be met.

Cases of such nature may be considered an oversight either by the contractor or a prescribed duty unknown to an individual in charge of the site or workplace.

If for any reason, the inspector finds more issues concerning health and safety that require his attention, additional verbal advice will then be issued on what is contrary to the requirements and how to perform corrections.

In some cases, the inspector may put this into writing, indicating non-compliance and several branches of the law and a copy would be given to the firm concerned.

• **Methods of Notice:** A site or workplace inspection that warrants a notice serving, signifies that the inspector has identified breaches of the law, and its level are hazards.

Here, the law requires immediate issuance of a particularly quick and effective means of implementing corrections at the site for non-compliance with Health and Safety at Work.

Officially, the issuance of this notice falls under two main categories; which are the **"Improvement Notice"** and **"Prohibition Notice".**

1. An Improvement Notice: this method of compliance is implemented only when an inspector who visited a site has found contravention of statutory provision and there is a higher possibility that the contravention will persist or be repeated.

An alternative could also be another case where the inspector witnessed the commission of this contravention at the time of visiting to the site unannounced.

The inspector would then issue an improvement notice requiring adjustment to workplace health and safety provisions stated and by a certain code of practice approved by the Health and Safety Commission relevant to the premises.

The law requires the inspector to include the following in the notice:

I. The particular location of the workplace or site, where the contravention occurred or found.

II. The specific provision contravened or uncompelling.

III. Detail on the reason by his professional opinion to which the said provisions have been contravened.

IV. A specific (duration) date in real-time by which all corrections needed to be performed or the matters are expected to be completed.

If the contractor or site manager objected to the Improvement Notice issued, an appeal can be made against the notice within a stipulated period of 21 days from the date of issuance.

This appeal has to be sent to an Industrial Tribunal for due consideration on the matter, and afterward, a consideration given may result in these three (3) with no exceptions **cancellation, confirmation,** or **modification** to the notice.

As soon as an appeal is lodged for further consideration, the requirements indicated in the notice are suspended awaiting the outcome from the appeal hearing.

2. A Prohibition Notice: As the name suggested, prohibition from the word "Prohibit" this compliance method places immediate suspension of workplace or site activity commonly known as **"Stop Work"** in many communities.

The inspector may consider issuing a **"Prohibition Notice"** when there is a health risk, usually, a serious personal injury occurring from the work activity, or an operation being carrying on or about to be carried out.

This notice made the prevention of any dangerous activity at the site possible, its emphasis discontinues of all of the unsafe work, (it is prohibited to do so).

Such operational function at the site ccould be carry out or continued when the requirements stated by the inspector in the prohibition notice have been met and any related contraventions resolved.

The issuance of prohibition notice can be found in used in two (2) ways;

Firstly, an immediate effect. This type of notice may be issued only when the inspector upon consideration anticipated that the risk of serious injury is imminent and any further delay allowing the work process to continue under such conditions jeopardizes human lives (worker(s) health and safety).

The notice will interfere with ongoing work activity, or an operational work procedure yet to begin with immediate effect till the required standards are met.

Secondly, prohibition notices of deferred effect. Even though the inspector has issued prohibition notices under deferred effect, the notice becomes effective from the date mentioned in the notice by the inspector during issuance.

From that specific date stated, deferred effect notice must be complied with, till the requirements are met before the prevention of work activity or operations can be resumed.

The efficacy of a prohibition notice ended, the moment contractor or site manager has resolved with the regulations of the statutes and full compliance implemented with all matters and any other contraventions.

The basic difference between Improvement Notice and Prohibition Notice are:

1. **With improvement Notices,** site work may continue during the lodging of an appeal via Industrial Tribunal by the contractor, or site manager and the outcome shall be complied with representing the final decision.

2. **Prohibition Notices,** however, are served by inspectors based on dangers anticipated, and when necessary.

The issuance of this notice by an inspector without regulations being contravened could therefore be a legal contravention that gave rise to the issuance of the said prohibition notice and all facts in particular, must be clearly stated or recorded in the notice.

SECTION II

Leadership In Process Safety Management (PSM)

Upon completion of this part, the HSE learners, professionals and trainers would gain thorough understanding of the basic knowledge required to build the framework of leadership in the process safety management.

This section also contains the fundamental approach of Health, Safety, and Environmental Management Systems in the Oil and Gas Industry.

Take-Away benefits of this part consist of the following elements:

- **Managing HSE Systems in Oil and Gas Industry**
- **Obligations in Process Safety Industry (PSI)**
- **Management of Change (MOC)**
- **Process Safety Management (PSM)**
- **Managing Contractors**

ELEMENT SIX

MANAGING HSE SYSTEMS IN OIL AND GAS INDUSTRY

Introduction

Over the years, the occurrences of incidents in the process industry have raised some series of concerns relating to how safety is being managed within various organizations in the petroleum industry and the global leading energy sectors and needs to reduce these preventable accidents of higher magnitude to the most acceptable or tolerable level.

Typically, about an inquiry into the deficiencies in leadership, and the anticipation of poor safety culture which is on the rise in different sizes of organisations that produces hydrocarbons in multiple geographical locations and improvements required to mitigate such defects towards a standardized practices suitable for an entity's present status and operations.

The growing of these questions requires correctional procedures to establish or re-establish effectiveness and efficiency in HSE management systems whose aim is to eliminate or mitigate all foreseeable causes of those major accidents to ensures both managing and working safely are attainable to championing the overall corporate success in the oil and gas management in upstream and downstream productions.

The activities of industrial undertaken where danger lies in all levels of their production cycles, such as voluminous of fires' poisonous gas release (loss of containment), hydrocarbon explosions, oil spillage, blowouts and others, occurring regardless of the company size that keeps questioning leaderships can't be ignored.

The severity of some of these accidents ranges from fatalities (loss of human lives) to huge financial losses, legal consequences, property damage, including environmental pollution such as land, Air, and water bodies and their habitats had become more alarming which requires managerial oversights improvements.

Finding effective control measures in the HSE Management Systems, especially in **Process industries**, the greater part of the attention was on the engineering solutions and design improvements that may be required due to the complexity nature of oil and gas installations.

However, some incidents in this Twenty-First (21st) century, involving hydrocarbon explosions such as Texas City in March **2005, US** and the Buncefield fire in December **2005, UK** signaled the red-flag that (continuous improvement in the PSM is required).

Whether or not **Engineering Solutions** and **Design Advancement** utilized previously wasn't the problem-solving facilitator, the Oil Spillage of **"Deep water Horizon"** also referred to as **(Gulf of Mexico Oil Spill)** in April **2010 which was** regarded as one of the largest environmental disasters in American's history speed up the **Search for Answers** across the Process Industry (PI).

Moving forward, the **leadership actions** then became a major contributing factor for scrutiny when inadequate leadership was identified for remedial action plans required and it should answer the following;

- With a lesser or no errors in oil and gas facilities, greater results in both elimination and mitigation controls for preventing such disasters in the petroleum industry could be achieve?

- With an increase in leadership knowledge over and above their current threshold in multiple hazards and risks level about their installations boosts their efficiency yielding a preventive mechanism to control major catastrophe in the oil and gas sector?
- Could the looming dangers surrounding the petrochemical production units and undertaken dealing with hydrocarbons production processes be mitigate?
- To what extent will leadership remedial policy last before another adverse event visits an oil and gas manufacturing company operating in Onshore and Offshore sites?

OBJECTIVE

The objective here, implies the required systematic directions in the process safety management (PSM) that deals with facts without allowing personal feelings to alternate the well-structured procedures created on the merit of due diligence in the process industry for the corporate entity's needs.

The primary objective of the Health, Safety and Environment (HSE) is to eliminate or mitigate work-related injuries against employees in the oil and gas industry using the Tool "Process Safety Management" (PSM).

To achieve operational excellence in dealing with hazards in refineries and petroleum products productions, the PSM plays a pivotal role in providing solutions to the companies in the oil and gas industry for the upstream and downstream management alongside the midstream activities.

Any well-structured PSM contains the comprehensive strategies designed to handle multiple hazards which rises throughout the petroleum industry worldwide.

PSM depicts the Risk Management Requirements designed to effectively manage the physical needs of these industrial processes such as the (Integrity of Operating Systems and Procedures for Handling Hazardous Substances).

Companies can apply for Adequate Design Principles and Improvements, Engineering Solutions, and Operating Practices suitable to their organizational needs.

The functionality of the PSM extended beyond the systems physical controls by looking through the level of actions required from the organizational leaders that would provide relevance to the Prevention of Accidental Release of Hydrocarbons.

MANAGING HSE SYSTEMS

The general intentions for practicing occupational safety and health, which factor into consideration the context of environmental management systems is relative to learning from previous case study and incidents.

Incidents that have occurred in varying industrial organizations and their impact on the environment are sometimes devastating especially, disaster experiences in the oil and gas industry.

At the core function of good practices on creating a safer workplace, the environment within which the organization's process plant has been established cannot be left out due to other stakeholders who could be affected by unsafe operational activities of the company.

When there is any catastrophe such as the release of methanol gases into the air or atmosphere, mass fire outbreak, collapse of oil rig, and so on.

Many equipment and machinery that are used in the process industries have been designed based on a high level of precision and accuracy to function in a certain way, and any malfunctioning of the process plant and maintenance failures could be detrimental.

Inadequate safety measures, failures in operating parameters by responsible personnel, and lack of attention to details could result in an unwanted outcome with incident leading to loss of human lives, damage to the company properties and environmental pollution.

This element, Managing HSE Systems in the oil and gas industry describes the approach needed to investigate incidents, accidents, near miss and dangerous occurrence caused by the operational activities involving hydrocarbon production.

The main goal is to establish two (2) basic requirements the (**Immediate Causes and Underlying Root Causes**) of incidents and lessons learned would help when making provisions for prevention that aims to eliminate or mitigate those scenarios.

This approach can be categorized into different steps that are suitable to fit the needs of the company in line with emergency response and disaster management, the fundamental theorem to begin with as a learner or safety professional, when you want to learn anything from a previous case study is by asking these questions:

Firstly, why investigating accidents and near-misses?

Secondly, should we use the same level of approach to investigate both?

Thirdly, Identify the incident category you are investigating and grouped them as follows:

1.Near miss

2. Accidents, Accident with injury/injuries, Accident with only damage to property, Accident involving both property damage and

injuries

3. Dangerous occurrence

4. Ill-health

Ideally, the next line of action should be:

I. Treat the injured first,

II. Secure or make safe,

III. Commence with the fundamental investigation procedures.

Four (4) or more different stages of approach can be deployed, where each part of the stage can be used to search for different answers that would simply addresses:

A. What went wrong?

B. What has already been done?

C. What needs to be done?

1. Method of Basic investigations

Stage 1. Gather Facts Information

Stage 2. Analyse the Information to determine:

- The Immediate Cause
- The Underlying Root Cause

Stage 3. Provide Suitable Corrective Measures

Stage 4. Plan for Remedial Actions.

Step 1. Information Gathering - this involves the process of obtaining all the required information about a particular incident. It has to fulfil the requirements below:

I. Secure the incident scene

II. Identify and interview witnesses' (do this one-by-one/separately)

III. Collect and record only factual information

IV. Make documentation reference

Step 2. Analyzing Information - this part would help in concluding the immediate and root causes of the incident. For example, an employee fell from the storage tank at 8 feet and suffers an Arm Fracture – causes are:

IMMEDIATE CAUSES

A. Deteriorated Ladder

B. Lack of inspection

C. Unsafe working conditions

D. Poor safety culture

UNDERLYING / ROOT CAUSE

1. Corrosive Ladder
2. Lack of Safe Systems of Work (SSOW)
3. Inadequate safety and health practices
4. Poor Maintenance

Step 3. Identify Control Measures - this is to ascertain the exact control measures suitable to deal with both immediate and underlying root causes resulting to the incident.

The aim of this action should fulfil the preventive or recurring which is best done following the safety Hierarchy of Controls listed below:

- Elimination, this is to remove the hazard from the work, As Low As Reasonably Practicable (ALARP)
- Engineering / Technical / Manufacturer controls, change the way people work
- Procedure / Administrative controls, isolate people from the hazard
- Introduce an appropriate Personnel Protective Equipment (PPE)

Step 4. Plan the Remedial Actions - this is the final stage where the mitigation plan provided needs to be rechecked to determine the implementation, level of effectiveness, and efficiency. At this point, you are required to provide:

I. Control Measures Recommended

II. Priority

III. Timescale

IV. Responsible individuals / Person

Below contains the cost of remedial actions and as well as an ongoing costs.

REMEDIAL COSTS	ONGOING COSTS
• Improve workplace safety	• Provide Permit-to-Work
• Ensure competence in health and safety	• Carrying out regular inspection
• Improve safety culture	• Provide Ongoing training for operators
• Increase preventive maintenance	• Maintaining the storage tanks with corrosive detection policies

ELEMENT SEVEN

OBLIGATIONS IN PROCESS INDUSTRY (PI)

The level of responsibilities required in the process safety management concerning organizations that work with or use hydrocarbons and others products' classified as Highly Hazardous Materials should be managed in conformance with all the safety-related processes of the industry at all levels without limitations.

The best practices of health, safety, and environmental management should be relative to the well-being of departments and other components that perform a function either internally or an external that are not permanently reside within the process entity should be treated with utmost importance.

For example, the safe transportation of hydrocarbons or any substances that are dangerous with the potential to cause harm to human health, property damage or capable to cause significant environmental impact should be handled according to the safety standard procedures to prevent loss of containment.

In this case, responsibilities become a requirement ranging from the managerial levels, departmental heads or line managers, supervisors, and through to other subordinates who can be used appropriately to ensure compliance with the regulatory requirements.

Strengthening the process industrial obligations should be made-to-measure the adherence of the company's health and safety policy designed for Onsite and Offsite safety, that's **(Onshore and Offshore**

safe operational envelopes).

The roles of all stakeholders must be designed and disseminated, including the accountability in line with a designated position apportioned to be handled by competent person with full-length of knowledge on hazard identification and risk assessment (HIRA) about the job.

This Knowledge base should also cover all the industrial risks that should be managed by the process safety, and the understanding of change in risks at each work stage, including the life cycle of equipment, plant or machinery used by the organization and different risks posed by them.

The need for competence in petrochemicals productions, especially the oil and gas industry, personnel should be embodied with multi-discipline risks assessment knowledge, such as new risks associated with;

1. Introduction of new equipment or machinery into the process,
2. Removal of existing equipment to increase or decrease productivity,
3. Technological changes caused by the removal, replacement, or addition of new plant.
4. Technological advancement to modify the existing product manufactured by the organization,
5. Modification to the existing plant used in production processes should be thoroughly assessed and new risks identified be managed as required by the management of change (MOC).

Given this, individuals such as directors, line managers, supervisors, safety practitioners, engineers, workers representative and others concerned, for example, community members should be actively engaged.

Through the appropriate medium of communication with detail information on hazard and risks and their respective roles in emergency situations and frequent updates where necessary.

The responsibilities of Process Safety Management, especially of substances described as highly hazardous materials such as hydrocarbon productions alongside safe storage and transportation requires the best practices at all levels and cannot be achieved by only the directors and top management.

Despite that, company directors and top officials reserves the overall responsibility and accountability to their subordinates, the complexity nature of oil and gas installations require different roles expected to be played at all levels such that directors and top management will delegates managers, supervisors, and workers in fulfilling them.

Assigning a responsibility of Process Safety Management (PSM) to a well-trained individual competent enough to perform their functions accordingly would help eliminate some level of human errors such as performing a task too late or too early in the work processes that could result in failure leading to a disaster.

All tasks should be clearly stated and communicated at all levels, from the organization's board, directors, managers, health and safety personnel etc.

Bring the Engineers onboard, includes the maintenance team or workers who look after the installations and performance day-to-day or regular activities can be helpful.

Acknowledging periodic maintenance personnel who might not be a permanent worker of plants and machinery used by the process industry, if any, provide them with the entity's safety policy.

It must be considered that, the road map for the corporation to achieve success in the process safety management represents **'Everyone Has a Role to Play'** therefore all those concerned are employed to do their part exactly as intended.

Understand the importance and urgency in managing hydrocarbons such that, one's failure could be the failure of all and does not matter the position held by such personnel in the organization.

The Board

The principles of Process Safety Leadership in the petrochemical industry, including companies that produces or uses chemicals classified as highly hazardous substances placed the following obligations on the shoulders of the industrial leaderships **(the board)**; it includes:

1. LEADERSHIP VISIBILITY, this requires the board member or members, or top officials of the entity carrying out site visits and making it regularly would be ideal, this way subordinates would see their physical presence serving as motivation emphasizing that, process safety management is the responsibility of all.

2. ASKING QUESTIONS, this is to show all stakeholders tasked with a specific function in the process safety management that, the company's board needs to know who does what, how, when and where?

3. LEADING BY EXAMPLE, here it depicts practicing or leaning by doing, for instance, the top management member or members wearing their personal protective equipment (PPE) during site visit, indicates high level of compliance with regulatory requirements affirming the rule which states that, no one is above the law.

4. ADHERENCE TO SITE RULES, this would clearly show the employees and others concerned that, all the written and communicated site policies and orders that must be complied with are still active, and their functioning is the reason why top management of the organization is following those rules during the visit.

5. PROVISIONS FOR RESOURCES, the board or management who form the corporate leaderships in oil and gas installations are expected to provide resources needed for the site and projects aiming to achieve the proper functioning of the process safety at all levels.

These resources may include financial resource, provision for qualified health and safety personnel for the plants, ensuring the availability of PPEs such as Masks, gloves, helmet, breathing apparatus and others will complement industrial success.

6. AIDING THE RISK ASSIGNMENT PROCESS, this point requires the board involvement in the risk assessment procedures either directly or indirectly.

Directly means their physical presence during the risk assessment processes to witness what has been identified to brief other board members.

Indirectly means no words or letter of interference from the board or top management to put on hold the ongoing risk assessment processes for the sake of continuous production, especially if the said risk assessment requires partial shutdown of the plant before the risk analysis can be conducted.

Risk assessment may also involve temporary shutdown of the whole plant when needed.

The motto of Process Safety Responsibility **'The role of everyone'.**

It should follow these simple details:

A. Everyone has a role to play in the process safety.

B. Such roles are required to be defined clearly with no limitations on communication.

C. Obligations must be allocated to people with the competence to carry them out.

D. Identifying those responsible to be held accountable for success or failure is a must.

For examples;

1. Managers: Do you part by allocating all the needed resources,
2. Engineers: Perform your role by designing and maintain plant, equipment or machinery,
3. Safety professionals: Issue all the necessary advice and guidance when need to ensure compliance,
4. Employees / Workers: Adhere to all the safety protocols according to the site policy without any personal feelings or believe.

STAGES OF RESPONSIBILITIES

This briefly described the category of responsibilities at different levels that must be observed in the process safety management (PSM).

At the Senior Leadership Level of the organization, below are the obligations placed on the Chief Executive Officer, (CEO) and others Leaders.

These above mention leaderships are expected to assure their respective firms have their readiness and preparedness to deliver competence in the management of hazards associated with the company's operational activities, they must:

1. Ask crucial question;
2. Ensure competence in the process safety management (PSM) at all levels, at all times;
3. Ensure continuity in development of expertise at all levels involving the process safety functions, typically with emerging of new laws, and technology;
4. Ensure the designed training are current, including scenario-based-training, and the medium used to train all personnel are accurate and updated;
5. Listen to the process experts, be it a complaint, suggestion, and experiences shared;
6. Ensure there is accuracy in the company's ability to manage or review the competence level of all the third parties or contractors whose failure could be a detrimental to your organization;
7. Establish suitable medium of communication to achieve effectiveness and efficiency at all levels, especially hazards identified must be communicated to the right recipient on time for necessary action, delay is dangerous.
8. Ensure the provisions for resources are adequate and must include time allocated for conducting risk assessment.

At the Senior Leadership Level, ensure resources which are appropriate should be made available, these resources that would be considered appropriate should include:

i. Human Resources such as Safety professional, primarily for advice and guidance to achieve compliance thereby keeping everyone safe.

ii. Financial Resource for the acquisition of materials and transportation to the site such as purchasing PPE, facial masks, gloves, helmet, breathing apparatus and others.

iii. Physical Resources like provision for first aid rooms and stuffed, make them accessible as designated area readily available to be use by qualified first aiders, when need.

CONTINUOUS IMPROVEMENT

The obligations of process industry (PI) require that, improvements should be made to existing PSM system in use. This is to meet with the organization's needs due to series of changes that may occur time to time caused by several factors.

Factors such as technological changes, change in law, modification to equipment or plant, removal of an existing machinery, or introduction of new ones and among others.

In short, the process safety leadership should set objectives and targets. This can be presented in four (4) basic parts below:

- **Establishing overall objectives** - What do the organization's leadership seeks to achieve?
- **Set Targets** - Decide on what to achieve first, by defining which one to begin with, (the stepping stones)
- **Monitor Progress of Indicators** - Identify what has been done to achieve success, and do not relent, but lagging for the causes of failures to remedy them.
- **Conduct Frequent Review** - Frequently Ask, what went wrong? If failed, what needs to be done, by whom, how, where and when, consider the appropriate time scale for regular review.

A Diagram showing Objectives and Targets In Process Safety Management

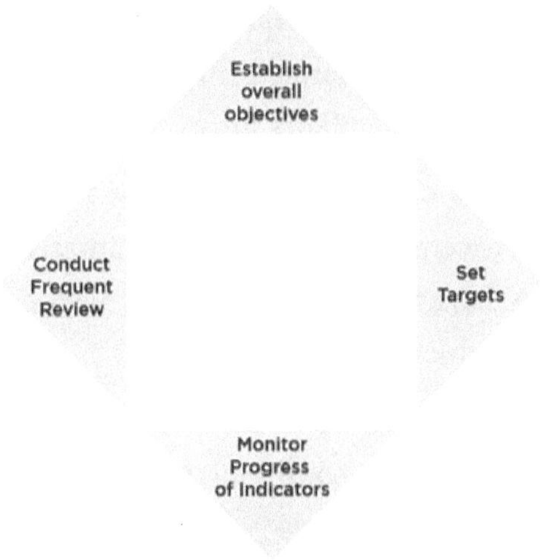

- Establish Overall Objectives concerning the process organisation
- Set Targets and utilise them as a stepping stone one after the other.
- Monitor Progress of Indicators using (Leading as success and lagging as failure)
- Conduct Frequent Review as regular measurement tool.

The establishment of a firm continuous improvement mechanisms for process safety management, leaders should not engage in only the active and reactive measures as an effective monitoring of the safety performance.

They can as well utilize the bench-marking methodology by comparing their company's process safety performance to that of another organization who uses the best practices in the same industry that has help them to reach a greater height.

The bench-marking principles could be achieved when they are utilized as progress indicators with an example shown in the diagram above.

Following all the prescribed requirements to identify hazards, and provides accurate mitigation mechanisms to protect all stakeholders that could be affected when hazardous substance is released is the form of bench-marking worth trying.

Companies develop plants, equipment, or machinery whose integrity may not remain the same; instead, it would be subjected to changes at every stage of the plant's life cycle while in used.

Such stages require different risk analysis, so as to the need for further safety improvements across the organization's operational activities and the hazards identified could be dealt with before causing any catastrophic event.

These and among others are the reasons behind the process safety improvements being continual process rather than being disheartening.

Other factors mentioned earlier could be technological change, with new standards of operations also called (safe/standard operating procedures /SOP's) that should be factored into consideration.

Without full consideration given to safe operating procedures in the process industry resulted in cases such as a catastrophic scenario that occurred in Bhopal, India in 1984/85 that claimed thousands of human lives.

The severity of that accident extended to devastating effect on the environment's health care facilities due to higher number of patients exposed to the poisonous gas that was released, and lack of adequate emergency procedures to respond to the affected community members prolonged the recovery thereby turned the crisis into a disaster.

ELEMENT EIGHT

MANAGEMENT OF CHANGE (MOC)

Introduction

Management of change abbreviated as MOC may be considered as a corporate initiative that should be handled by only top management or high-ranking personnel in an organization.

But in truth, work and operational routines are affected by it every day, and all the affected employees need to know the change and new hazards or risks introduced into their work processes.

By the requirements, individuals who are affected by the change directly need to understand all the new procedures of work due to the change brought by the MOC for their safety, the safety of others, and the environment.

Code of Federal Regulations (CFR) such as Safety and Environmental Management Systems; SEMS - 30 CFR Sub-section 250 off-Shore industry, OSHA-29 CFR 110.19 of Process Safety Management Standardization, and so on require compliance with the use of management of change by chemical industrial operators.

MOC must be applied to every type of change regardless of the size except **"Change In-Kind"** which are performed often by normal and operational routine technicians or maintenance team in the process facilities.

To ensure efficiency in the functioning of employee safety and health, many process industrial organisations including the Oil and Gas production units uses the management of change approach as the best method to protect workers, their assets, the general public, and the environment as a whole.

Introducing any form of change can produce unexpected results, therefore when you decide to do things differently, then it must be done right to ensure it doesn't lead to preventable accidents.

The fundamental rules of management of change give you the systematic approach on how to do this change.

Most of our industrial accidents occur because of the failure to identify the hazards created by a change and provide measures for adequate controls.

Management of change (MOC) has a golden rule which states that, for every change that occurs, normal operations cannot be continued until all hazards associated with that change are identified and analyzed.

Followed by adequate control measures being introduced to eliminate or mitigate to safeguard employee's safety and that of others who could be affected by the change in the process.

Typically, in the process safety management (PSM), any change no matter how small it is can adequately introduce new complexity into an existing complex process in the oil and gas installations and chemical manufacturing establishment, including pharmaceutical companies.

As change can be considered as one of the most constant things in life, it must be managed carefully in all working environments, especially in companies that utilize or manufacture materials of hazardous nature.

Such chemicals are harmful and have the potential to cause ill-health to humans and other organisms when in contact, or exposed.

Companies especially in the process industry that deals with hydrocarbons such as refineries, oil and gas exploration and petroleum product storage facilities are required to conduct their operational activities in compliance with all the statutory requirements.

This is to ensure that all stakeholders who can be affected by the work of the organization such as workers, contractors, vendors, suppliers, members of the society have been protected.

This also includes the required safety parameters to cover where the process plant has been established, alongside the processes to safeguard the entire environment from pollution by preventing the release of harmful containment into the Air or Atmosphere, and Water bodies.

Management of change when implemented systematically, would have the coverage of land pollution mitigation measures that may arise from oil spill in connection to the activities of oil and gas undertaken and such good practices can be advanced further through suitable environmental management and sustainability projects such as the ISO 14001:2015 EMS standards that need to be considered.

Management of Change

Management of Change or (MOC) is a systematic approach that represents the best practice, it serves as a requirement often used in the process industry when the systems installed needs some changes due to multiple factors.

Management of change is meant to ensure that, the safety and health of workers and other stakeholders who may be affected by the undertaken's activity has been safeguarded due to the recent change performed on the installations.

The implementation of MOC considers that, risks on the environment are adequately controlled when an organization makes changes relating to their facilities such as operational procedures and processes, changes to instrumentations, change in production output be it an increase or a decrease.

Other coverage includes change in documentations and equipment with the sole aim to safeguard all from potential harm, before, during and after a transition process brought by the changes made.

Management of change practices can be introduced by changes in process technology, changes to machinery, designing new products using existing plant through research and development, equipment downturn affecting production capacity and so on.

The change in process technology may cause by changes in the manufacturing rates resulting from market high demands of the entity's commodity, removal of existing equipment or an introduction of a new machinery.

Additional reasons include product advancement or development of new commodity due to lack of raw materials, experiments and texting, damage to equipment part, adjustment to safety critical element, or a plant malfunctioning (deviation from designed intent).

Management of change could either be a permanent or temporary, but to avert any catastrophe, it remains the best practice to ensure all hazards associated with the changes in the process have been identified, reviewed, analyzed and control measures needed to eliminate or mitigate are established.

In workplaces where highly hazardous chemicals are used or produced, the Process Safety Management or (PSM) regulations must work together with the management of change to ensure things are adequate handled before resuming operational activities to keep everyone safe, especially the affected employees.

The MOC Prescribed by OSHA

The United States Occupational Safety and Health Administration (OSHA), clearly described the need for MOC before implementing significant changes in a company's operational procedures or part of the process that would be affected by the change.

To ensure workforce safety when managing organizational changes, the Management of Change, OSHA 1910. 119(1) stated that except for **"replacement in kind"**, the employer shall establish and implement a written methodology to manage changes to process chemicals, equipment, technology, procedures, and changes to facilitate that area of the process affected by the change introduced.

The OSHA 1910. 119(1), prescribed the structural requirements for adequate implementation of management of change which covers the process of evaluating and recommending change, with clear explanation typically in the refineries and petroleum companies in areas of:

A. Process Safety Information (PSI),

B. Process Hazards Analysis (PHA),

C. Operating Procedure,

D. Mechanical Integrity (MI), and

E. Management of Change (MOC).

For an ideal MOC, the Occupational Safety and Health Administration (OSHA) recommends the requirements listed below:

I. Consider the Design of Equipment; the coverage of equipment design should include,

1. The installation of control valve bypass,

2. The installation of an oil spillage guard berm under a fracturing tank or a storage facility together with suitable grounding and bounding,

3. Ensuring changes in alarm set point, and

4. Other changes to materials of construction.

II. Operating Procedure changes the following are required,

1. Changes in procedures for the manually controlled addition of methanol to a chloride injection tank or storage facility, and

2. The installation procedures for any new type of relief devices that are different from the original.

III. Procedural changes in Inspection, Testing, and Maintenance should include,

1. Introduce changes in the previous inspection intervals or timing for piping circuits,

2. Consider the needed changes in the number of thickness measurement locations with an alternative of monitoring location for conditions assessment,

3. Adherence to maintenance changes procedures that follow any form of change with exception of Replacement-in-kind to process equipment.

IV. Changes in Facilities must include

1. The installation of a light wood or metal alternative to shed structure near a Hydro-cracker Unit, and

2. The change needed for the control room to be located within the process safety management coverage process unit.

V. Consider the Limitations of Time, typically to all Temporary Changes. This part requires the level of consideration that should be given to the existing temporary changes that would prevent the initiation of management of change.

VI. This final part emphasizes the proponent of MOC initiation where the process plant has existing temporary changes until it has been completed and closeout.

This must include all temporary changes initiated as support during installation of new vessels or piping circuits, or using shed.

It can also be a broken part of the control room area as a temporal measure during repairs or maintenance or the construction main control room.

Temporary changes that are often initiated during the ongoing permanent change being made should also be considered for a thorough risk assessment.

Giving enough consideration to the following are equally helpful,

1. The workflow processes including a document of MOC such as (Forms and Processes involved)
2. The Responsibility and Authority Needed to Implement Change in the organization
3. The Levels of Approval requirements to be met before commencing

the MOC process.

Establishing effective management of change processes is highly recommended as a cornerstone of safety for organizations in the process industry.

It depicts the formal documented system developed to identify any needs for modifications.

Any properly designed MOC follows hazards identification and risk assessment methods to ensure that, the length of implications of any change in the company are well understood by those affected by it either directly or indirectly.

This could help guarantee that, new hazards and risks introduced by the change are identified in addition to the existing ones to avoid any unwittingly increased.

The core function of MOC process should be aligned with the review of any proposed modifications using the existing operating parameters and design criteria.

Additionally, for any new plant and equipment installed, the management of change processes should be used as a tool to evaluate and record any planned changes to all the safety-critical devices.

Factors to consider should include Replacement of Equipment with Non-identical alternatives, changes to alarms, and changes to other operating parameters, specifically those which do not include the **'Safe Operating Envelope'.**

Documents to support the MOC Application

The appropriate development of management of change and implementation requires the involvement of the following documents to support the application procedures.

- An authentic or original process design criteria,
- Drawings of an existing process or part affected by the MOC implementation.
- Accurate information about the proposed changes, which should include, all the affected Electrical and Mechanical Equipment Specifications
- Detail information of trip and alarms planned
- Risk Assessments, and
- Hazard and Operability Studies/Analysis (HAZOP's)

The safety term HAZOP is a risk assessment tool designed systemically to be used by the process industrial organizations to identify and address any potential hazards before it causes an incident that could affect the safety and health of employees and other people, causing damage to the company's assets while hindering productivity.

The Application of MOC

Ideally, management of change implementation must conform with the points listed below:

- Clear definition and communication- the MOC must be clearly stated in written and communicated to all those in the approval processes.
- Timely submission-the proposed management of change must be submitted in advance by the individual proposing the modification to all stakeholders who are relevant to the process.
- Include the concern units such as the Engineering department/ engineers, Health and Safety department, manager or personnel, Manufacturing Department, Research and Development, Unit Operational Staff, Maintenance team or personnel and so on.

- Follow up and managed as the proposed changes pass through the hands of different stakeholders' reviews that come with approval, disapproval, or make comments.
- Obtaining the MOC final approval from suitable and appropriate individuals responsible for authorizing such significant changes to the plant.

In addition, formal approvals are often required to be granted by the member or members of top management for highly significant changes, for example, the removal of safety-critical devices from the process.

Documenting MOC Changes

Except for change or replacement-in-kind, all changes made to the process plant and process design should be documented correctly to help guarantee process knowledge retention, as such it should relate to the following:

1. The management of change files should document categorically consisting of all the changes proposed, approved, implemented, and records retained based on the formal procedure for continuous improvement and other purposes.
2. Any documents classified under the process, like process and instrumentation diagrams, operation manuals, and others should be updated accordingly.
3. The MOC file retained should contain the current modification or design specifications for further use as future reference.

Consult and Inform Stakeholders affected by the change.

In the process of consulting and informing all parties affected by the changes, there is a need to consider the following:

A. Understand the process industrial needs that, effective change management are best achieved resulting from collective efforts of teamwork and decision-making, likewise effective consultation instead of the decision made by one individual.

B. The best practice of MOC required all those affected should be consulted through the process, upon completion and the final changes made should be communicated.

Conducting (MOC) Training for the affected Personnel.

Safe Operating Procedures relative to effective change management require that all the staff affected by the changes needs to be trained adequately for the MOC.

If the management of change is necessary, then before obtaining MOC approval, the coverage of MOC training should be extended to those in the approval processes for clearer understanding and knowledge to ensure competence at all levels.

Case Study

Management of Change- the Flixborough Disaster in 1974, United Kingdom, UK.

Flixborough Incident, 1974, Image 1

Flixborough Incident, 1974, Image 2

Flixborough Incident, 1974, Image 3

The picture of the above disaster which occurred in the UK in the year 1974, introduced a significant highlight on the need for effective management of change regardless of how little it could be, with an exception of replacement-in-kind, which is considered usually as a routine maintenance.

This empirical scenario and others have claimed human lives in numbers. This particular accident is a typical case of human error that killed about 28 people, damaged properties, fire, explosion, and environmental pollution.

Details: The accident originated from a proposed management of change, (MOC) which was considered as a modification to pipe work that aims to allow maintenance on a leaking vessel.

The tasked was performed without conventional reflection on the original design requirements and was unfortunately carried out by people who lack the needed competence and this resulted in the fitting of a wrong pipeline being inefficient to resist the amount of pressure that it experienced in the process plant.

That pipeline got ruptured, which led to the release and explosion that caused 28 human lives and according to the official inquiry conducted by the authorities, the following observations were made in the report:

1. No calculations were performed to determine whether the bellows or pipe can resist this distress or pressure,
2. No reference was made to the relevant British Standard prescribed by regulators or any other accepted standard,
3. No reference was done to the designer's guidance issued by the manufacturers of the bellows,
4. No drawings of pipe were made, instead chalk was used on the floor of the workshop,
5. No pressure testing was conducted on either of the pipes or the complete assembly was made before it was fitted.

Due to these, the assemble as constructed was of completely an unknown strength and without any of references relative to the British Standards.

In the subject matter, Chemical Process Safety: Learning from case histories such changes are described as **"modifications made with good intentions"** and there are several scenarios of such cases where despite the motive of trying to do the right thing for the process plant, with an inadequate and accurate level of hazard evaluation, risk assessment, and planning, the outcome was undesirable.

These included tanks collapsing on the road and tanker trucks falling, reactors exploding, with no rational thought did the management of the organization intend to do the wrong thing, yet the outcome was disastrous.

Management of Change Benefits

Firstly, it is required because, the MOC when effectively designed and implemented, would help reduce an unplanned adverse impact on the company's system's integrity, security, stability, as well as reliability for the firm's procedures being altered or added.

Secondly, the management of change increases and improves the employee's productivity and efficiency in the areas of planning, coordinating, and implementing changes.

Thirdly, it provides stability in the organization's production through the provision of safer working environment for the employees and other stakeholders who can be harmed in the event of a disaster.

Fourthly, Management of change is needed for proper hazards identification, analysis, and control measures to eliminate if possible, or mitigate including emergency response procedures.

Lastly, it is required to be used as the best practice for both Temporary and Permanent Changes in an organization to avoid any failure that could result in catastrophic scenario.

ELEMENT NINE

PROCESS SAFETY MANAGEMENT (PSM)

Introduction

Viewing the general operational activities in the process industry that are relative to the oil and gas productions in global energy sectors at all levels requires a well-designed process safety management system (PSMS) that are flawless to;

- Provide employees safety,
- Promote public health and safety,
- Protect company's assets,
- Protecting the environment,
- Provide returns-on-investment, and
- Promote compliance with regulatory requirements.

The Process Safety Management or PSM provides the fundamental mechanisms required to fulfilling the industrial safety needs in relation to the petroleum production companies such as oil refineries, offshore drilling rigs, onshore natural gas processing plant, hydrocarbon storage facilities, crude oil transportation and safe disposal of industrial wastes.

The core-function of process safety management is to serve as the shield of preventive needs in remedying the widespread of safety failures in the process industries, and to represent an internationally recognized standards of best practices in a complete package form which oil and gas companies can tailor-made to suit their needs.

Analyzing the structural framework of every good designed PSMS, their purpose defines a unified system of multiple components in units that works in unity at a specified unit of different process entities to unite all of its functions to replace industrial interdependencies with interconnectedness inputs towards sustainable systems integration in-line with components compatibility analysis, would pre-sets process safety management (PSM) as an ideal good fit for fixing future foreseeable failures faced by the oil and gas undertakings who manages complex portfolios within the upstream and downstream production of hydrocarbons.

The PSM In Focus

In generic terms, the PSM can be used to describe a well-structured safety management system compromise of multiple risk mitigation methods that are designed differently but used in combinations for different purposes under a single umbrella towards common goal of achieving a safer working environment within complex installations in the process industry whose operational activities relates to multiple hazards managements' which can be classified as a high-risk-ventures.

The suitable designs of process safety management systems (PSMS) should have the coverage of an unrelated-working-together-elements such as, management of change (MOC), fire and explosives safety management, emergency response procedures, environmental protection designs, hazard identification, multiple risk assessment tools alongside safe systems of work (SSOW), plant management and assets integrity monitoring, including the oil and gas installations start-up and shutdowns etc.

The context of process safety management in this book however, concerns itself with the level of safety in the general management of Hydrocarbons Production Cycles that must be safe at all levels such that, industrial preparations of petroleum products should be handled using a well-defined approach representing the required systematic procedures suitable for an organization in the process industry.

The PSM in this case, acknowledge corporate entities that work with, or produces hazardous substances that losing control of such chemical containment could be a detrimental to the company itself, the employees, members of the public and significant environmental impact on other non-human life forms.

To achieve industrial excellence in upstream and downstream production processes of hydrocarbons, this book contains the Golden Rules in Process Safety Management required for oil and gas companies to excel in both managing and working safely in their installations.

The complexity of day-to-day operations of the process industrial organizations in different regions globally, especially those in the oil and gas sectors that have to manage multiple functions in their facilities which are interconnected and any failure could result in a devastating effect and to prevent this from happening, requires a robust PSM to facilitate safe operations at all times.

PSM Failures and Consequences

Can PSM fail? Incident-1

The consequences of events in the past arising from varying defects such as **human failures,** unsupervised and poor maintenance work, mechanical failures, unidentified installation's equipment deviating from design intent and the inappropriate uses of technological machinery are best mitigated by process safety management.

The appropriate implementations of the content in this PSM book will foster the creation of competent professionals to aid the possible elimination or mitigations of; (lack of repairs, failure to detect and predict unsafe working conditions at their early stages, failure to provide the needed safety mechanisms that ruined human lives, causing property damage, polluting the environment and non-compliance with legal requirements).

Additionally, the proper functioning of a specific designed structures of process safety management systems can derive maximum efficiency when considerations are given to the cause and effect of incidents in the process industry.

The illustration below, is an example of failures in the process safety management systems due to poor supervision which justifies an empirical of human errors.

PSM Failures due to poor supervision

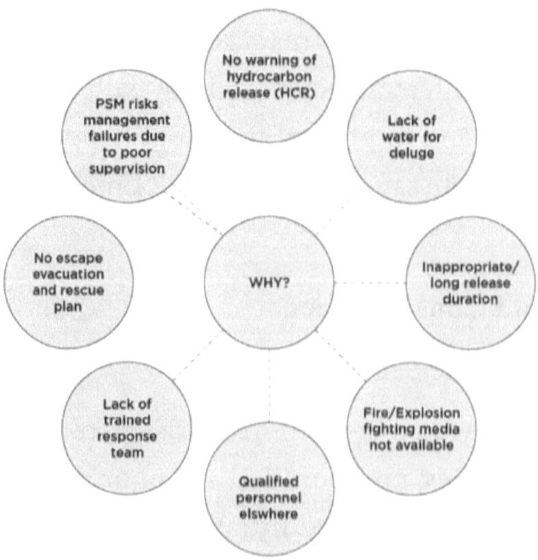

Learning from the above diagram, the PSM failures that occurred identifies human error as the culprit, indicating an accident that could have been prevented when the PSMS are not being undermined and the required competent personnel assigned with the task execution.

The Causes and Effects

How can PSM fail? Incident-2

The structural design of PSM that considers cause and effect analysis as an input, should include factors like, the capabilities of the process machinery and equipment performance, their efficiency and reliability because having detail knowledge about plant's lifespan-lifecycle-maintenance-records would help oil and gas companies to determine how risks are been managed.

The following diagram is an example of PSMS failures due to a failed Risk-Management, one of the most important components of process safety management systems and this incident is another preventable case scenario.

PSM risks management failures and causes

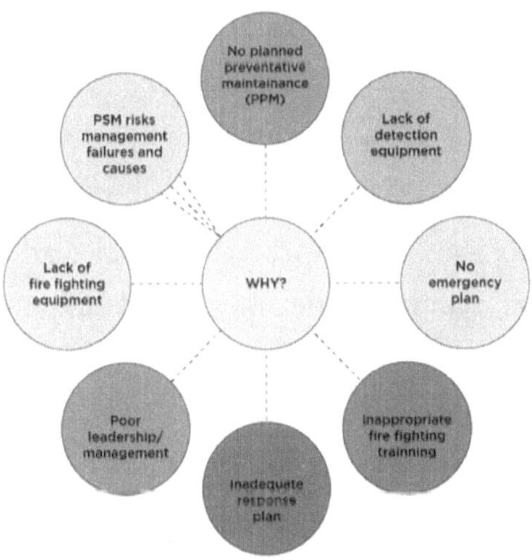

Therefore, relating the PSMS designs for only the **oil and gas installation's** Causes and Effects is a direct call to action that requires the implementation of control recommendations that are active for prevention instead of reactive after an incident have occurred.

Hydrocarbons production management includes complex installations, multiple hazards and high-risks such that, PSM covers fire and explosive methods of classification, should be managed by Modern Risk Management Systems and each specified design must be inclusive.

This way, process safety management in oil and gas risk-management to match Cause and Effect will be a specified **Failure Mode and Effect Analysis (FMEA)** that are comprehensive risk assessment technique with a designed coverage for analyzing **all-causes** and **all-effects** alongside the Extent to which the Effects can impact the Oil and Gas Facility in question.

The Duo Index

This Duo index is the representation of the author's work in reference to the industrial recommendations in the process safety management (PSM) made by two (2) of the pioneers in respect of Dow and Mond designed protocols of INDEXES for the safe management of fire and explosion hazards alongside chemical compound toxicity exposures in the process industry (PI).

According to standards, these industrial leaderships upon the creation of their separate indexes, utilized different approaches to categorize both hazards and risk levels faced by the process entities and below methodologies of Fire and Explosion and Toxicity were used as an acceptable risk classifications technique.

Seeking efficiency in **Hazard Identification and Risk Assessment**

(HIRA) to further differentiate between fire, explosions and chemical compound toxicity associated with process industrial undertakings, the author will take you back in time in references to tow (2) different good works titled;

a. Dow Fire and Explosion Hazard Index, and

b. Mond Fire and Explosion and Toxicity Index.

I• DOW FIRE and EXPLOSION HAZARD INDEX, this was designed by Dow Chemical Company, an American multinational corporation that specializes in the manufacturing of chemicals fertilizers and commodities for agricultural industrial applications, including the production of plastic products.

Being in an operation for more than 100 years, Herbert Henry Dow's company developed a method used to rank the relative fire and explosion and their risk associates in the process industry.

Designed by Dow Chemical Company, this analysis are based on the systematic calculations of numerous hazards and explosion indexes, subjected to the material characteristics as well as the process data components.

Dow fire and explosion hazards management approach can be shortened as DF & EI (Dow Fire and Explosion Index).

II• MOND FIRE and EXPLOSION and TOXICITY INDEX, this approach uses similar features of DF&EI except that, more effort would be extended to the toxicity of the chemical substance in question and through the development stages of the process plant.

Be unspecified application to safety professional, Mond approach did not ranked the toxicity of hazard separately, instead it was based on the general information about the inherent risk factor of hazard exposures in the process industry.

Those approach could be used by non-specialized safety personnel, and process development team based on techniques used to analyze the cost of all inherent hazards being considered as one of the major factors in the selection and evaluation processes of the plant.

Mond fire and explosion and toxicity risk management design can also be referred to as (MFETI), Mond Fire and Explosion and Toxicity Index.

Summarizing **the duo index** based on the similarities between DF&EI and MFETI while utilizing "TOXICITY" as the differentiator, the following risks were classified according to their methods, and could be utilized in the process industries and industrial chemical plant with a lesser or no hydrocarbons production management needs.

Their indexes coverage could be applied as Risk Estimations "Numerical Index", Risk Classifications "Level of Risk", and Risk Ranking with Alphabetical Codes, also termed as "Grade", bringing the author's conclusion on **the duo index** being presented to you as the tabular representation below.

Δ Dow Fire and Explosion Hazard Index (DF&EI)

Δ Mond Fire and Explosion and Toxicity Index (MFETI)

THE DUO INDEX RISK CLASSIFICATION TABLE

Index	Level of Risk	Grade
• 1-60	LIGHT	E
• 61-96	MODERATE	D
• 97-127	INTERMEDIATE	C
• 128-158	HEAVY	B
• 159 +	SEVERE	A

Below are the interpretations of the above given risk classifications, use it as a practical function of Dow Index.

Give a specific feature for each of the factors in the fire or explosion hazard analysis and simplified them as:

1• Material Factor (MF) as the main items

2• Process Hazard Factor *General Categories* (F1)

3• Process Hazard Factor *Special Categories* (F2)

4• Hazards Factor in all Process Units, formulations as F3 (F3 = F1 + F2)

5• Fire and Explosion Index (F&EI) F3 x MF

Below are the merits and demerits of DOW and MOND fire, explosion and toxicity

Merits	Demerits
Reproducible, it can be reproduced or by copy	Meant to be used at the early stages by designers
Proper function needs the evaluation of every part of process plant and units with potential to increase the severity of fire or explosion are identified	Difficult to apply to plant that already exist, since core purpose of early stage has passed
Easy to apply or use on plant prior to their commission	Lack of efficiency when used to determining normal expected loss to the process plant
Limitations, this application is based on existing / real fact experience	Interpretation needs judgement
Acknowledged by industry as a good practice for Chemical Process Risks Ranking	

Additional PSM control measures

Apart from the need to address fire and explosion in process industrial organizations, the following control measures can be utilized to improve the overall fire safety performance.

I• Inventory Control: this has to do with the required resources such as raw materials, and intermediate goods used, as well as the finished commodities produced by the process entity that should be allowed in stock depending on their nature.

For example, a chemical process plant that manufacture chemical paint for construction use, these are flammable and the availability of such finished products in the facility should be regulated in the inventory at all times.

II• Distancing the Operating Plant: this point described the creation of spaces between the operating plant thereby separating them from each other.

III• Positioning and protection of control room facility for the process environment, and the isolation of critical equipment

IV• Provide Suitable Administrative Controls, for instance, Management of Change (MOC). Management of change controls are required to oversee all activities different from normal in the process plant.

Examples are:

1• Carrying out a replacement activity on a plant, equipment or machinery with non-identical parts. This is also known as Not-In-Kind Replacement.

It may occur when, the process wants to increase production rate above the current output, decrease productivity below the present production rate, developing new products that requires an alternation to existing plant and so on.

2• Introducing new plant or machinery and removal of existing equipment from the process operational activities

3• Considering making changes to the existing operating procedures, (Here, if it is external, then ensure to established the design basis in addition to process Safe Operating Envelope. Management of Change Coverage: The MoC processes should be conducted formally and must identify:

A• The Scope of the Change - this part needs to identify what needs to be changed relative to the proposed change, and which part of the process is affected by it be it partial, or the entire unit.

Detail understanding for the needed change are established and communicated to all stakeholders required to be notified

B• The Accountable and Responsible Personnel - Identify all personnel involved in managing and making the changes.

Ensure a clearer roles and responsibilities has been assigned and communicated to those tasked to undertake the modification procedure in the facility.

(Prevent gabs) by mapping the organizational changes

(Further training) by matching personnel and their skill sets to the task

(Phase the Changes) by dividing the change into several phases, for example, from phase 1 - 3

C• Identify how risk analysis is been carried out, for example, before, during and after the MoC has been completed.

Understand the existing risks before the change, and any additional risks introduced by the change.

D• Channel of Communications - Identify the various channels used to communicate the changes to all the stakeholders.

Inform those affected by the change to ensure everyone understands how to perform their duties assigned in the function.

E• Identify Competence - View this in two separate ways,

Firstly, training, skills, knowledge and experience of personnel involved in managing and making the change, and

Secondly, training needed to be provided to all personnel whose work is affected by the change, and who is the competent individual to deliver such training.

Ask how, when, where and what is included in training and through which medium?

For example, face to face in the facility's training room, different shifts making it impossible to access all personnel at the same time, therefore online is preferred.

Number of personnel attended, how well do they understand the change communicated to them, and the new way to conduct their routine tasks assigned.

The MoC Basic Documentations: Below documents may involve in the management of change procedure under (PSM) for proper evaluation;

Firstly, Process Flow Diagrams (PFD) - this diagram is used commonly by chemicals and process engineering to show the flow of plant processes and equipment.

Its core function depicts the existing major equipment in the plant facility and how they inter-relate, and may not show minor items such as piping and instrumentation diagrams (P&ID's) and their designations.

Secondly, Basis of Design (BOD's) also called (Original Process System Designs) - this design is a document that shows decisions behind a major design made to meet with Owner's Project Requirements (OPR).

The owner's project requirements in this case are meant to include the list of process information, such as why these particular systems were selected and designed, the client or owner's expectations from the process plant and so on.

Thirdly, Equipment Specifications - this part deals with the specifics of plant and machinery, their functions, manufacturing brand, their lifespan, operating risks, productivity and others

Fourthly, List of Controls - this document should cover all the control system of the plant such as alarms, trip settings and so on

Fifthly, Cause and Effect diagrams - this document intends to show a potential cause of a specified event and their effects on the process entity.

The basic function of the diagram is to outlined the product design and provide quality defects control mechanisms to identify the potential sources causing the overall effect.

Such causes are generally categorized into various groups to help identify and classify them as the major sources of variation.

Lastly, Drawings of Hazardous Areas - this document is need to ascertain the detail classifications of all units that are hazardous in the plant.

This part, hazards and their types are shown together with their risk's associates.

MoC proposed document: this document describes the design elements and issues related to the process plant which management of change is to be carried out. The change proposal documents are as follows:

I• The proposed change and its description

II• Realistic date of the proposal

III• Specific reasons for the intended change proposed. Item required to be shown included, why the proposed change is needed.

For example, (modification to the existing plant to increase the production capacity above it current limit due to high rise in demand for the entity's commodity 'P' in the market)

IV• Different personnel to authorized different kinds of change.

V• Bring on board all the competent personnel to assess the MoC process

VI• Monitor to ensure personnel adhere to all the change protocols

VII• Include an independent auditing of Safety Management System abbreviated as (SMS).

This may help serve a purpose for determining if the process operators are using the current required functions of an international standards, for example, certified by ISOs such as (ISO 45001), (ISO 14001) and others.

ELEMENT TEN

MANAGING CONTRACTORS

Introduction

Contractors' management become a need and is based on the complexity nature of the petroleum industry, part of the company's operations is sometimes allocated to contractors due to several factors.

Factors such as risk transference, higher level of expertise required for the job, high cost of machinery acquisition needed for the task, expensive equipment maintenance or repairs, long duration or time required to train personnel to become qualified for the job and others.

To manage another organization, (A Contractor) who has been awarded either a short-term or long-term contract to perform a specific task for an oil and gas company should be handled with utmost importance.

To ensure part of the whole complex tasks is undertaken by a firm who possess the competence, experience, technical expertise, resources and good safety records to deliver in order to eliminate or mitigate disaster of incident, can be achieved by series of selection processes.

Regardless of the contractor's status, their relationship with the oil and gas company they worked for and the specified task assigned to them, any of their failures that may results in catastrophe would have a significant impact on the main company.

MANAGING CONTRACTORS IN THE PETROLEUM INDUSTRY

Who is a contractor?

A contractor is this context refers an individual, a group of people or an established organization who are professionals and capable to provide skills, services or practical and technical job they are tasked with by their employers.

In a broader view, there are different types of contractors who undertakes varying degrees of contracts from people and companies for a set period such as hours, days, weeks, months or even years and any certain time frame given is dependent on the project to which the contractor is needed.

A contractor who does not form part of the ownership of the business they worked for, may be referred to as the "Second Party" in law terminology, and may range from Self-employed contractor, an independent Sole trader or sole proprietorship, or a limited liability company for large scale contracts.

Understanding contractors without any specifications could be the used for a person or an organization hired to work on contract basis for an individual or a business entity.

Describing contractors and their activities with no specifics however, can be categorized under the following three (3) basics.

I• Independent Contractors: These types of contractors are professionals specializes in an industrial specification with little or no multi-task function rather than their area of specialty such that, independent contractors are hired to execute a specific task in line with their expertise.

For example, Saudi Arabian Aramco oil company contracted Japanese Chiyoda Kako Kensetsu Kabushiki-Kaisha popularly known as **"Chiyoda Corporation"** to build oil refinery and gas process plant.

Headquartered in Yokohama Japan, Chiyoda Corporation is a well-known engineering firm in the global energy market, with an outstanding specialization.

They are expert in the designing and construction of oil and gas midstream facilities for gas processing and Liquefied Natural Gas (LNG), downstream refineries and other petrochemical installations.

An independent contractor of this nature, would only be engaged to undertake the design and construction of oil refinery and gas processing plant, while serving the engineering works and maintenance.

II• Subcontractors: These are professional individuals or entities tasked in a contract with a responsibility focus on fulfilling only a given portion of an ongoing projects.

A subcontractor in this case is usually hired by the main company (contractor) to whom the whole contract or project has been awarded.

In some cases, a short-term direct hiring by the project owner would suffice when the task assigned to the main contractor does not include the portion executed by the subcontractor.

Most of the functions performed by subcontractors are found within or as part of a whole project being handled by another contracting firm executing the project major, (PM).

III• General Contractors: These are professional bodies of people or an organization tasked with full responsibility for managing the whole construction projects, usually from the start to finish.

Under this category, contracting firm is solely responsible for the management of the entire construction works being the Project Major (PM) assigned according to the contract agreement.

Often, companies of this nature can perform multi-task functions during the execution of the contract as long as the purpose of those multiple tasks forms part of the project being carried out by the company.

MANAGING CONTRACTORS

Despite the fact that, companies operating in the oil and gas industry have to manage different contractors for different purposes to fulfil their industrial needs, there should be a formal approach designed according to Industrial Standardization due to high level of hazards, and their inherent nature of risks associated with the production of petroleum products.

Among others, the scale of contractors used by organizations operating in the petroleum industry include:

- Support Vessels,
- Diving Services,
- Drilling and Exploration rigs operational activities and so on.

To maintain safe working with external contractors, consider to:

- Perform risk assessment on all tasks assigned to contractors, (Risk assessed their job)
- Ensure contractors and their employees abide by the site rules in line with the company's Health and Safety Policy
- Establish suitable medium of communication. Define appropriate communication channels should ensure that, the workforce of contractors know the identity of site personnel to

contact in times of emergency and how to contact them.

- Make provisions to provide contractors with a well-structured Safe Working Procedures and should include Permit-To-Work systems (PTW).

However, to achieve success in contractors' management, as well as working safely with contractors the following systematic approach would be appropriate when utilized.

The fundamental steps for managing contractors.

1• Planning

2• Contractor Selection Process

3• Design Systems for Contractors working On-Site

4• Keep Monitoring System Functions

5• Consider to Review their Work

A• PLANNING

It is thc fundamental requirements to have a well-designed plan in place when an organization decides to engage external contractors for their industrial nccds. The planning part should include the following:

I• Identify the need to engage external contractor with the expertise to execute a specified task

II• Determine the length of duration for the contractor's job. For example, a state-owned oil and gas company in Dubai called Emirates National Oil Company (ENOC) operated by Dubai Government has awarded 1year contract to Japanese firm, Chiyoda Corporation to provide engineering services and maintenance on its newest refinery and gas processing plant built 2 years ago.

III• Provide a direct line to supervise all tasks undertaken by external company with contractual obligations to your organization

IV• Conducting background checks on the contracting company

V• Ensure to know more about the contractor's performance, strength and weaknesses from their previous clients, (employers)

VI• Consulting with group members or head of contractor's association for findings

VII• Engage in a Specialized Consultations from external source, if necessary.

B• CONTRACTOR SELECTION PROCESS

It is an ideal for oil and gas companies to consider the design and implementation of a procedural structure for choosing the right contractors for the job.

Such structure should contain all standards of best practices that could be used to justify reasons for the contracted company selected from the pool of multiple firms who applied for that specific job.

In this case, procedures below could be additional value to the organization when used appropriately:

1• Contractor's Health and Safety Policy (identify their compliance with legal requirements)

2• Education, training and experiences gained by the contractor's employees

3• Contractor's years of experience in the industry

4• Obtain detail information about their last 3 projects executed, if not available by the contractor, consult their former employers (at least the last three or more)

5• Demand for any specific certifications from a certified body, for example, ISO 14001, ISO 45001, and others

6• Look for any recent enforcement action taken against them by industrial regulators

7• Determine the contractor's level of competence. For instance, number of;

Failed projects and reasons,

Projects executed successfully under normal conditions,

Projects executed with difficulties like engaging sub-contractors.

C• SYSTEMS FOR CONTRACTORS ON-SITE

Designing systems of work for external contractors on-site is a useful tool, for example, begin daily operations from 0900 hours, 30 minutes after the permanent employees of the oil and gas company have begun their daily tasks at 0830 hours.

The designed systems should consist of a set of rules, and site policies outlining the coordination of day-to-day operational activities of contractors, that should be integrated into the existing function of the company's site operations.

This point would help introduce the share believe of working together from different angles, different task purposes but seeking to achieve the same goals and objectives for the process company in the petroleum industry.

Systems for contractors working on-site should involve:

I• Adherence to health and safety policy,

II• Obtaining permit-to-work for high-risk tasks

III• Following the establish communication channels for reporting incidents, accident and near misses to concern department and personnel

IV• Following the safe evacuation protocols during emergency

V• Using the required safety tools and equipment. For example, utilizing the prescribed Breathing Apparatus, Personal Protective Equipment (PPE) etc.

Using the above-mentioned requirements to manage contractors, should be accompanied by establishing procedures for monitoring to ensure all designed systems are being implemented at all levels within the organization.

Afterwards, the final consideration should be the schedules to review contractor's work, support your techniques by conducting, Contractor's Competence Assessment (the CCA).

Factors should include;

- Years of experienced gained in their type of work?
- Training acquired in offshore/onshore safety requirements?
- Level of compliance with regulatory requirements, and
- Identify if any enforcement action has been taken against them in recent times for non-compliance and why?

Establish a Safe Shift Handover Methods that must be followed at all times (detail knowledge of hazards or work-related issues should be included) using:

- The provision to allow enough time for Shift Handover,
- Convey all meetings to be held in a formal manner,
- Create appropriate method to deal with Conflicts Arising Between the shifts,

- Determine the length of how much the out-going shift achieved, in actual results and in real terms,
- Provide a medium to handle - other matters arising, for example, the out-going shift left some work they don't want to do?
- Consider to include contractor needs to achieve a safer working environment.

SECTION III

Plant Management And Process Industry General Operations

Upon successful completion of the third part of this training material, the content of the elements would help learners and safety practitioners to demonstrate understanding and core competence in technical applications with comprehensive knowledge of familiar and unfamiliar conditions in the process industry's plant management and its operations in general.

A specific take home opportunities cover the areas of:

- **Plant Operations, Inspection and Maintenance**
- **Installations Start Up and Shut Down**
- **Permit-to-Work Procedures and Coordination**
- **Essentials of Safe Shift Handover**
- **Floating Production Storage and Offloading (FPSO)**

ELEMENT ELEVEN

PLANT OPERATIONS, INSPECTION AND MAINTENANCE

Introduction

The general concept of conventional maintenance of plant systems in the oil and gas industry is to support the optimal operations of plants, equipment and machinery in their day-to-day functionalities.

The goal is to establish accurate production capacity and improve the quality of outputs while safeguarding employees from harm, injury and occupational ill-health.

However, the benefits of proper maintenance can also be extended to the reduction in the likelihood and severity of industrial machinery and plant unscheduled downtime, while fostering a decline in the overall operating costs within the industrial processes.

PLANT OPERATIONS

In generic terms, plant operations and maintenance (PO&M) defines all the functionalities that ensure the safe, adequate and efficient use of plants and equipment by operators to provide health and safety to workers, members of the public, the environment and the organization's assets.

In line with modern-day technological advancement, several methods could be utilized by companies in the oil and gas industry to achieve maximum efficiency of plants.

Methodologies such as Asset Integrity, Inspection, Planned Preventative Maintenance, Intuitive Maintenance Modules, Accurate Reporting, Maintenance and Compliance, Well-Documented, Corrosion Control Procedures and among others are essential to the firms operating in the petrochemical industry.

Asset Integrity- It defines the capability of an 'asset' to perform its intended function effectively and efficiently across the estimated useful period according to the design intent.

Here, the planned life cycle of the asset should correlate with its ability to safeguard the safety and health of individuals exposed, alongside the protection of the operating environment. The fundamental life span of an asset should follow the steps below:

NO.	Lifecycle Phases of an Asset	Deliverables or Activity Samples
i	Design Models	Safety Studies
ii	Construction and Hook-up	Procurement Quality Standard Plans
iii	Commission	Functionality Test
iv	Operation	Maintenance / Repairs
v	Modification	Management of Change (MoC) Process
vi	Decommission	Decommissioning Plan

Inspection and Services - Activities such as monitoring, testing and inspections are often conducted by legal requirements. For example, testing pipelines for corrosion, lifting equipment and inspection for corrosion and among others shown in the table blow.

Testing Pipelines for Corrosion and...	Lifting Equipment & Inspection for Corrosion
Post-installation / Re-installation	Before being commission for the first time / after refitting to prevent faults from incorrect installation.
At the event of deterioration causing significant risk	Items of an equipment left out in all weathers / being exposed to chemicals or bad environment (harsh weather conditions)
At the event of exceptional situations may jeopardize safety	After major modifications or repairs has been carried out. After known or suspected damage, change of use or after long time of disuse.

The inspection and testing services on Pressure Vessels, Pipelines, should be conducted following a written or well-documented procedural examinations and sometimes such tests are carried out to meet the requirements of regulatory bodies.

Safety measures of pressure vessels should be conducted by the internal crew members, external inspection teams or both.

Unless the company's policy states otherwise by special arrangement, for instance, the schedules of hydraulic testing to be conducted every 5 years could be mandatory as a legal prerequisite.

Lifting equipment schedule for every six months inspection by internal personnel or 12 months by external inspectors should be done formally and in written procedures, unless specified differently in writing by the operator.

Planned Preventative Maintenance (PPM) - This is the systematic approach designed and utilizes in the form of pre-planning arrangements aimed to provide prevention to the company's assets from breaking down before they can be repaired and the term "PPM" defines any maintenance process that is organized, documented and scheduled dependent on either Condition-Monitoring, Time-based or Both.

Planned preventative maintenance also known as Scheduled Maintenance, however, is a supervision mechanism used to combat an unplanned downtime of machinery, plant or equipment when there is the availability of Personnel, Spare Parts and a well-defined strategy to make use of these resources.

Time-based Preventative Maintenance: - This type of maintenance activity describes a calendar-based timetable pattern where companies choose to carry out the needed maintenance works on their existing equipment or machinery in operation that are mostly categorized under the execution of a pre-set frequency for that asset.

Even though there are some level of variations in equipment and their functions, business entities in the petroleum industry can schedule such maintenance frequencies based on these multiple factors.

i. Manufacturer's Instructions,

ii. Warranties,

iii. Organizational Policy by Experience, and

iv. Statutory requirement.

For example, complying with the HSE regulatory bodies at national and international levels serves adherence to the prescribed HSE Laws which could benefit the process entity's compliance demonstration records.

Common example of time-based maintenance is the Quarterly (3monthly) inspections and servicing of Fire Extinguishers pressure gauges and expiry dates, Fire Control Panels, Fire Sprinkler Systems, alongside Smoke and Heart detectors.

Condition Monitored Preventative Maintenance: - This defines a specified maintenance technique focuses on predicting when equipment failure may occur and remedying such from happening, adequate maintenance is performed.

Assets specified under this group are the equipment that can communicate or transmit their real-time status or condition to the programmed maintenance system of the facility or installation.

In pursuit of best practices in occupational safety and health, process industries can adopt the use of progressive technology and integrate them into their HSE systems to further strengthen efficiency and effectiveness.

Companies in the oil and gas industry can acquire an **Integrated Workplace Management System (IWMS),** a software platform suitable to administer and optimize these resources allocated and readily available at the site for the organization to achieve maximum usage from its equipment and plant, and other infrastructural assets.

Furthermore, software system like **Computer Aided Facility Management (CAFM),** can help enable site managers to plan and execute tasks and monitor all operational maintenance involving reactive or accurate reporting and planned preventative maintenance or the utilization of intuitive maintenance modules.

Additionally, a C**omputerized Maintenance Management System (CMMS),** is software built to facilitate asset management by scheduling maintenance activities while tracking work orders for progress reports.

Efficient use of these monitoring methods provides just-in-time maintenance that can reduce the costs of maintenance and prevents equipment or asset failure in connection to the reaction maintenance processes.

For example, if the temperature of the installation exceeds its limited values, immediate preventative maintenance required is performed to prevent it from failing.

MAINTENANCE TECHNIQUE

1• Shutdown Maintenance - This type of maintenance can only be performed when the said machinery has been out of service or switched off. Shutting down equipment may be costly to the organization though, sometimes, a damaged component to be replaced or the maintenance activity on the plant requires a shutdown as the only viable procedure to ensure personnel safety.

2• Emergency Maintenance – This kind of maintenance is carried out when the company's asset or equipment part breakdown or changes in condition during its function within the operational activities with the potential threat to employees' health and safety or could result in property damage that would be a detrimental

3• Opportunity Maintenance - Preventive maintenance is classified as an opportunity when a task such as the replacement of equipment component or machinery parts are fixed under "convenient".

This convenience justifies any maintenance work that took advantage of the planned or unplanned shutdown of the plant system and resources required for the maintenance is readily available at the workplace.

Resources in place such as maintenance personnel, an item for replacement, and among others should be suitable for the tasks, and sometimes, personnel utilize the opportunity to perform working adjustments that couldn't be made when systems are in operation.

TYPES OF CORROSION AND CONTROLS

Corrosion occurs as a result of the deterioration of a material (metal) and its critical properties caused by either chemicals, electrochemical, dissimilar metal or material contact, and other exposure reactions of the material surface from the surrounding environment.

Some common forms of corrosion and control measures.

NO.	TYPES OF CORROSION	CORROSION CONTROLS
1	Uniform Corrosion (General Wastage Material)	Monitoring, Inspection and Testing for Corrosion
2	Stress Corrosion	Cathodic Protection
3	Dissimilar Metals in contact (Galvanic Corrosion)	Surface/Protective Coatings
4	Hydrogen Embrittlement	Corrosion Inhibitors (Chemical treatment in pipelines)
5	Fatigue Corrosion	Initial Design Consideration
6	Inter-granular Corrosion	Selection of Material (Choose inherent corrosive resistant)
7	Localized Attack (Pitting)	Environmental Controls and Process

METHODS OF SYSTEM MONITORING

The general approach to monitor systems in the process industries are the three (3) described below:

- **In-line Systems;** This monitoring procedure utilizes integrated operational processes in which devices are placed directly into equipment or machinery have to be removed for continuous data collection and data information obtained becomes a real-time monitoring of corrosion like bio-studies, coupons, probes and weight-loss.

- **On-line Techniques;** This methodology involves corrosion monitoring on periodic basis using a fixed devices like sensory or instrumentation in the process equipment and systems to verify the parameters.

- **Off-line Monitoring;** This approach requires non-destructive testing and inspection for data collection and analysis outside the equipment operational processes with the following being common examples; Data for Laboratory Analysis, Manual Inspections and so on.

ELEMENT TWELVE

INSTALLATIONS START-UP AND SHUT-DOWN

Introduction

Start-up and Shutdown methods are essential procedures in the process industrial organizations, including petrochemical production units, typically of Upstream and Downstream management in oil and gas.

They define all of the systematic approaches to ensure the safe and efficient initiation and discontinuation of operations in terms of machinery, systems, or equipment in an installation.

Commissioning and decommissioning differ between entities due to the nature of their equipment used and commodities produced, they must be carried out under strict supervision in line with safe or standard operating procedures.

Start-up Techniques

For a successful start-up typically of hydrocarbon-containing equipment, a permit-to-work system is required to control such activity. It must cover the following:

1• Pre-Start Checks - Before commencing the equipment and systems installed, conduct the pre-install examination to ensure that all the needed conditions are met with no limitations.

Factors to consider should include appropriate critical safety functions, intact valve positioning, and required fluids, other lubricants like hydraulics and machinery oil gauges are functioning and ensure to verify power sources.

2• **System Arrangements** - Prepare equipment, instrumentation, and systems for start-ups that could be made to follow the guidelines issued by the manufacturer and other conditions such as Safe/ Standard Operating Procedures (SOPs).

It should cover establishing the Pressure or Temperature adequacy, purging air from the systems, pump priming, and others.

3• **Sequence Control:** This represents the start-up part that must be arranged to follow the sequence pattern in an orderly manner and correctly without deviation, examples may include ramping up power gradually or increasing energy input on sequence intervals.

4• **Functionality Verification** - Here, process industries are required to run tests on the equipment and systems while in operation and this is to ascertain adequate performance in machinery components and instruments.

5• **Hydrates Presence** - Hydrates and their presence describe crystalline structures in solid form created when Hydrocarbons combine with water molecules relative to specific pressure and temperature conditions.

This solid form of crystalline structures when generated can block valves, pipelines, and others typically under-deep-water environments or sub-sea with prevailing high pressures and low temperatures.

Shut-down Methods

I• **Inform Personnel:** Before shutting down machinery or systems, notify all the affected workers through suitable communication channels and follow up to verify, include shut-down details, and define responsibilities to relevant employees. For example, maintenance personnel, operators, engineers, and among others.

II• Shutdown Sequence: A brief description of reversing the start-up sequence such as closing valves, orderly manner for deactivating system controls, utilizing shutdown sequence per manufacturer's design, and a safe shutting down of an Auxiliary Equipment.

III• Energy or Power Disconnection: It clarifies the safe removal of power and energy inputs to the machinery from their sources like, shutting off the electricity supply, isolating multiple sources of fuel, and the closure of other energy supplies.

IV• Post-Shutdown Review: Conduct maintenance or inspections if needed, then extend to ensure that, equipment parts are secured, adequate safety systems are on-site, no fuel leakages, and valves are closed.

V• Draining/Depressurization: This category falls under the Working Safely methodologies in the HSE Systems and just as the name suggested the term "HSE" without any specifics prescribed, should fulfill Health, Safety, and Environment office.

In this context of safety, however, the working safely techniques should be applied, coaching by Managing Safely where safely depressurize the equipment and systems while allowing valves and pressure to dissipate.

WATER DRAINING PROCEDURES

The general function at this stage represents the need for insight into systems and other units of the processing facility and ensure to find the outlined procedural functions for water removal and separation below:

Firstly, Separation – The production steams in the oil and gas industry utilizes processes through Separation Equipment, commonly called the **"Separators"** to separate water from hydrocarbons.

An alternative name Hydro-cyclones can be utilized to carry out these separation processes such that, both devices use differences between density and gravitational forces to separate the phases.

Secondly, Water Knockout Drums - These drums are vessels designed based which enables water to settle out through the manufacturing stream by the force of gravity, and the separated water would then remove from the vessel's bottom

Thirdly, Gas Dehydration Unit - The gas dehydration units serve essential components that are widely used by Natural Gas Processing Plant and their function techniques involve the draining of water from natural gas or crude oil.

Absorption utilizing glycol-based solutions, membrane separation, and other common forms such as Desiccants Adsorption processes, examples include the use of Silica Gel, Molecular Sieves etc.

Lastly, Electrostatic Coalescer - This involves the use of electric field energy to the coalescence of Water Droplets, enabling them to combine and separate, utilizing this combination and separation process aimed to increase the efficiency of removal from the hydrocarbons phase.

COMMISSIONING AND DECOMMISSIONING

The conventional function of commissioning and decommissioning remain two integral parts in the life-cycle of the process industrial facilities, which include undertakings operating in the oil and gas industry.

COMMISSIONING

The act of commissioning describes the approach used in a systematic order aiming to ensure that, new installations and systems or modified equipment and machinery are fully functional and readily available for efficient and safe operations in the facility.

Commissioning procedures should include:

1• **Pre-Commissioning;** this includes a set of activities like checks, testing, and inspections to verify the installation completeness and accuracy that meets the prescribed quality of industrial standards.

The facility needs should cover instrument calibration, precise mechanical completion, functional tests, and equipment examination to ensure the overall systems have been installed correctly.

2• **Commissioning Planning;** this involved a comprehensive commissioning plan created to show the Scope, procedures, place, objectives, and required resources for the commissioning activity including, identify the sequence of activities, and establish safety perimeters while defining testing protocols for the installations.

3• **Safety Tests;** All the required perimeter testing must be conducted to ensure adequate safety systems, reliable fire protection mechanisms, emergency shutdown devices, and other safety-related specifics are functioning and matches compliance requirements.

4• **Performance and Functionality Testing;** Conduct itemized function testing on each component individually, subsystems, and the overall integrated system, check and verified communication interfaces, interlocks, and control logic are properly operable.

Initiate performance testing that focuses on assessing the system or facility's operational performance while considering the monitoring tests for Response Time, Efficiency, Accuracy, and Capacity in line with performance parameters.

Conclude by utilizing the results obtained, compared against industry standards, design criteria, specifications are in conformance to the general design intent.

5• Training Requirements; Ideally, commissioning should factor in training provisions sufficient for personnel such as operators, maintenance teams, troubleshooting, and maintenance needs for new machinery, equipment, or systems commissioned while assuring competent personnel on site, thereby promoting their capabilities to operate the facility and installations safely and efficiently.

DECOMMISSIONING

Decommission defines the set of arrangements made for a safe shutdown, removing and retiring equipment, systems, or machinery services from the installed facilities when their estimated lifespan has reached.

Decommissioning processes are as follows:

I• Planning - A decommissioning planning should consider stakeholders engagement, safety measures, assessing environmental impact and adherence to regulatory bodies when developing decommissioning plan.

II• Shutdown and Isolation - Ensure machinery, equipment and facilities have been shut down safely provide a safe disconnecting to electrical power, closure of valves, tag-out or lockout methods in place, isolate other energy sources including fuels.

III• Decontamination - These procedures are carried out to ensure safe removal hazardous substances, and materials while neutralizing any other containment available in the facility and it should cover the flashing, cleaning and adequate disposal of hazardous chemicals.

IV• Dismantling and Removal - The safe process of disassembling and removal of equipment, system and installations and such activities involve proper controlled of dismantling, cutting or demolishing other physical structures and machinery as well as safeguarding waste management and recycling.

V• Restore Site - The site restoration activity defines any condition where a decommissioned site could be kept for reuse- in-future-processes such as soil erosion mitigations and clean up exercises, restoring living organisms. Ensure compliance are met at all levels.

ELEMENT THIRTEEN

PERMIT TO WORK SYSTEMS AND COORDINATIONS

Permits-To-Work, abbreviated as PTW, can be classified as a specialized nature of safe systems of work used in all sectors to effectively and efficiently manage complex risks to ensure workplace tasks that are hazardous has been performed safely.

The motive behind PTW systems is to ensure that, specific workplace activities are prohibited or restricted and would only be allowed once stringent checks have been conducted by an authorized person because, permit-to-work procedures implies a specified risk management technique required to be design by personnel with expertise and the needed competence.

It represents a formal document in work processes which consists of careful examination of workplace tasks with the aim of identifying all hazards, and their risks associates and provide suitable mitigations to remedy them, including assessment on all safety elements in line with safe systems of work of the oil and gas installation to provide the needed systematic approach to perform that particular task safely.

Successful utilization of permit to work systems and coordinations are meant to ensure that non-routine activities of high-risk nature are planned, risk-assessed, authorized and carried out on an intended manner to helps minimize or eliminate human errors, typically those of different knowledge, and assumptions that are incorrect and mixed messages.

The fundamental features of any good Permit-to-Work document is used to define a task within the working environment that is being carried out, or performing or about to take place with a permission given by authorized personnel competent do so.

The general characteristics of a standardized PTW should contain the steps below:

1• The permit title and permit number, for example, (Permit Title, Modification of Plant ABC) with Permit Number (ABC-PL-006-2024-11/MDF)

2• References to other permits in connection to the work. This is also known as the Complementary Permits, that must be referenced.

For example, (Isolation Certificates Required to be in place to indicate the number of systems (equipment) that has been isolated prior to the task at hand.

3• Job Description, such as (performing a hot metal welding activity near hydrocarbon under pressure storage tank in oil refinery)

4• Job location or Plant identification including Equipment or Machinery, Circuit, Distribution Board, and others.

5• Hazards Identified in connection to the task, and the necessary precautionary measures taken for safe execution of the job.

6• Identified all protective equipment required to be used by personnel involving the task execution, and should include the Personal Protective Equipment (PPE's)

7• Authorization - this describes the approval obtained from personnel with the competence to authorize the permit issued for that particular task to commence.

8• Date, Time and Duration - this part of the permit should indicate the date and time the permit was issued and the time period the permit allows the personnel to work on the assigned task.

9• Identification of Employees - here, the workers who are tasked to execute the job should be mentioned, for example, the Electric Welder, Fire watch, Supervisor and so on - remember to record their given name(s)

10• Permit Acceptance - this indicates the name of personnel who accepted the permit and has crossed check if all requirements have been met

11• Extension of permits - these occur when the task to which the permit was issued is incomplete within the specified duration given on the permit and requires additional time period for completion.

This part may also include the handover arrangements such as the Shift Handover or Change of Shift

12• Returning to service upon successful completion of work - this section is to clarify that; workplace has been left intact and in a safe condition for normal operations to resume in the oil and gas facility.

13• Permit Cancellations - there are several factors that could lead to the cancellation of an issued permits.

For example, the permit indicates that, 5 employees are carrying out the job using grinding equipment B, and F-type drilling machines, but 7 workers were found, grinding device A, and G-type drilling equipment are being used, implying a deviation from the permission authorized by the permit.

ROLES AND RESPONSIBILITIES OF PERSONNEL

The roles and responsibilities of those utilizing permits-to-work should clearly defines according to a specified category, because different individuals with competence are required to be involved and not just a single person to cover up the whole process.

A well-structured PTW procedures is best achieved through the steps which are typically used to represent a specific employee assigned with a direct role and responsibility in the process.

It should be conducted as such to attained effectiveness and efficiency through collaborative efforts without bypassing any part of the approach unless the permit is meant to address trivial issues about work.

Permits-to-Work roles and responsibilities should conform with at least:

- the Permit Applicant,
- the Issuing Authority,
- the Performing Authority, and
- the Permit User(s).

I• Permit Applicant - permit-to-work applicant is represented by the personnel who requested for the issuance of permit to cover a task to be performed outside the company's routine operational activities that are hazardous and requires additional safety control measures under stringent supervision.

II• Issuing Authority - this part represents competent personnel tasked with the permit preparation and it should define the work to be carried out, identifies all the hazards involved and determine the appropriate safety measures taken as precautions to control them.

Such individuals must be familiarize with workplace hazards based on Job Risk Analysis (JRA) and the permit content should include the following;

(1) the exact nature of the task,

(2) the exact location where the work can take place,

(3) the names of each employees or personnel authorized to execute the task,

(4) date and time that particular work can commence,

(5) the valid duration of the permit in question,

(6) the needed control measures that must be introduced before, during and after the work,

(7) any other restrictions required to be implemented, and

(8) additional permits that may be relevant to the work. Such permits can also be referred to as the Complementary Permits.

III• Performing Authority - this section represents individual employees nominated and assessed by Head of Department (HOD's) or Supervisor as Performing Authorities with their final approvals obtained from the Operating Authority, typically in the Offshore Oil and Gas facilities, the Offshore Installation Manager(s) (OIM) and Production Manager(s).

Their roles and responsibilities cover the following;

A• Producing relevant work permit for the execution of targeted task and assumes an active role as a team member for conducting risk assessments and other processes

B• Making provisions for site visit prior to the execution of tasks as part of the hazard identification process

C• Make sure that, all complementary permits, plans and isolation certificates relevant to the job are complete and accurate where necessary before the commencement of the task

D• Facilitating Tool Box Talk meetings aiming to ensure that, all concerned parties are involved.

For example, Health and Safety Supervisor, Equipment or Machinery operators (Area Operator), and gaining all the affected work team members to understand the scope

E• Ensuring that, only the task covered by the scope of the permit-to-work (PTW) is being carried out, including the competence of employees authorized to perform a function on the job.

This section includes additional training required to be performed for personnel tasked to carry out the work where necessary and communication processes.

F• Make sure that, worksite is kept safe and in a tidy condition during the execution of task and upon the completion.

IV• Permit User(s) - this part of the permit concerns with individuals who are actively involved utilizing the prescribed job of the PTW.

Permit users are required to comply with the systematic approach stated in the permit alongside the paramount consideration for required training and competence in the use of systems, it must include;

- Specified tools or equipment that should be used,
- Personnel to handle that mentioned tool,
- Adherence to the time frame given,
- Adequate monitoring to check effectiveness,
- Enforcing restrictions of unauthorized personnel to the working area, and
- Providing feedback on PTW job assigned during the execution where necessary for determining efficiency.

ELEMENT FOURTEEN

ESSENTIALS OF SAFE SHIFT HANDOVER

At the core function of any good practices in occupational safety and health, the effectiveness of handling shifts between multiple teams tasked to perform certain responsibilities at the workplace should base on fundamental requirements as a standard procedure.

The ultimate goal of this method is to accurately share task information relevant to the work process in the oil and gas installation and it should be conducted through reliable communication channels across the various shift changes or between the outgoing and incoming personnel.

Taking into consideration the context of HSE, the effective handing over shifts across multiple teams performing routine tasks should be relevant to some level of the undertakings basic function, it should consist of at least the following three (3) principles:

1• **Preparation Period** - this part involves the set of arrangements to be made by the outgoing team or personnel before the end of their shift.

2• **Information Exchange / Handover**- this section defines the handover communication process where all the necessary information relevant to the job is passed on from the outgoing team to the incoming personnel to ensure the task-related message are shared properly to fulfilling work safe processes and task continuity.

3• Verification of information- this part describes the responsibility of the incoming crew to Cross-check the information received from the outgoing personnel immediately after they assumed task responsibility at the site.

Despite the differences in task routines, when a shift is ended personnel leaving their shifts are required to hand over all tasks and responsibilities to the next employees, (the incoming team).

The effective handover principles should include a report of all incidents or accidents that have occurred during the last shift process as well as any other outstanding tasks or relevant issues in connection to the work that needs to be addressed by the person who took over the shift.

COMMUNICATION PROBLEMS IN SHIFT HANDOVER

I• Long duration of maintenance work; -Where maintenance on plant, machinery, or equipment is being carried out within multiple shifts, or maintenance works that require more than one shift period in the work process before it can be completed.

II• Changes in system function; - The interruption of a machine's functioning on an automatic operation that needs to be switched to manual mode. An example of such a condition is where safety systems may be overridden. Fire systems such as deluge need to be switched onto manual function.

III• Deviation from design intent; - Such a situation could be attributed to an equipment function deviating from normal working. For instance, plant or machinery breakdowns.

IV• Absenteeism; - When extended absences of employees (a worker didn't show-up)

V• Lack of competent personnel; - Here, handover difficulties may occur between experienced and inexperienced workers.

EFFECTIVE HANDOVER MEASURES

A• **Efficient Communication** - Establish and ensure effective handover communications at all times and makes it a priority

B• **Design efficient handover procedures** - Ensure handovers are conducted face-to-face

C• **Adequacy in 2-way methods** - Ensure handover techniques embodied two-way procedures such that, both shift crews (outgoing team and incoming personnel) take joint or complementary responsibility for information accuracy

D• **Time Sufficiency** - Provide adequate time frame for the shift handover process

E• **Multiple communication channels** - Allow the use of both oral or verbal and written communication to share handover information

F• **Consider** analyzing the information needs of the incoming personnel

ELEMENT FIFTEEN

FLOATING PRODUCTION STORAGE AND OFFLOADING (FPSO)

What is FPSO?

In the Oil and Gas industry, the term FPSO is used as an abbreviation for Floating Production Storage and Offloading, and the acronym FPSO, defines a type of floating vessel used in the oil and gas industry primarily for offshore production, storage, and offloading of hydrocarbons.

Often, companies in the oil and gas sector employ FPSO specifically for deepwater or certain remote locations where the processing and storage of hydrocarbons is economically unfeasible by the construction of fixed oil platforms.

The design of the FPSO and its floating nature makes it ideal for mobility purposes and can move from one remote location to another thereby providing the flexible solutions needed for oil and gas production in various offshore fields within the deep-water environment.

The key feature of FPSO is that it possesses the ability to float on the surface of the water, utilizes its processing facilities for treating and separating oil, gas, and water, and store the produced hydrocarbons after extraction using the storage tanks before they are emptied onto the tanker vessels for transportation to the various refineries and other distribution facilities.

FPSO plays a critical role in our modern oil and gas industrial facilities specifically in the development of offshore drilling and reserves found remotely or in deepwater environments and is widely used as the most common and effective method for processing hydrocarbons and the extraction of oil and gas.

FPSOs provide organizations in the oil and gas industry with flexible mobility that allows them to be relocated to different fields as required making them cost-effective solutions for global oil and gas production and ideal versatile for varying degrees of geological and economic situations.

Breakdown of Acronym FPSO

The oil and gas industrial term FPSO which stands for Floating Production Storage and Offloading can be explained further as the breakdown of the abbreviation, and they are as follows:

F - Floating: the designed standards of FPSOs allow the vessels the ability to float on the water's surface which allows them to be moored at deepwater offshore sites where fixed oil drilling platforms or structures might be inoperable.

P - Production: FPSO vessels are equipped with an extraction facility to practically extract, process, and treat both oil and gas underwater reservoirs, including the effective separation of oil, gas, and water while serving additional necessary processing steps as well.

S - Storage: the designed structure of FPSOs has storage tanks onboard enabling the holding of the processed hydrocarbons for safe storage and those tanks are capable of storing crude oil until the arrival of the tanker vessels to offload and transport them to the refineries.

O - Offloading: FPSO drilling vessels are equipped with offloading mechanisms like underwater pipelines also known as loading arms. These loading arms can facilitate the transfer of all the stored hydrocarbons onto pipelines or shuttle tankers for transport to the organization's onshore facilities.

FPSO Ghana Kwame Nkrumah MV21

FPSO Ghana John Agyekum Kuffour

Fundamental Textual Representation of FPSO

From the diagram above, the following textual representations could be used to describe the FPSO for further understanding of its general overview with the basic systems and key components:

I. Hull and Mooring Designed Systems:

1. In generic terms, the FPSOs are built on a big buoyant hull that aims to provide stability and further buoyancy.

2. Mooring lines are included in the FPSO designs with secured anchors for effective operations at offshore locations.

II. Upper part Facilities:

1. Processing Units: the FPSOs contain processing facilities to treat the extracted hydrocarbons, including the separation, dehydration, and other processes.

2. Production Facilities: these production units are inbuilt equipment of the FPSO or the extraction of oil and gas from underwater reservoirs.

3. Gas Compression and Treatment Units: these facilities helped compress and treat the produced gas.

4. Water Treatment Facilities: FPSOs are equipped with systems for separating the produced water, as well as their treatment

5. Power Generation Systems: the design of FPSO contains Power Supply Generators necessary for the supply of the required electricity to the vessel.

III. Dwelling Units:

1. The FPSOs are designed with living quarters to facilitate accommodation modules for the ship's crew members

2. These living quarters contain dining facilities, accommodation rooms, toilets, bathrooms, and sometimes recreational areas.

IV. Helideck Sections:

1. The topside facilities of FPSOs are equipped with a helipad which enables a helicopter to transport to and from both incoming and outgoing personnel (the shift crew).

2. The helipad can also be used to facilitate airlift rescue activities in times of emergency

V. Flare Tower Compartment:

1. These sections of an FPSO are used specifically for the burning of excess gas in a more suitable and controlled manner.

VI. Turret Systems and Riser:

1. Turrets are the designed mechanisms that allow the FPSO to move around its mooring point in a rotation manner.

2. Risers are used to describe the vertically located pipes that enable the transport of oil, gas, and water between the FPSO and Seabed

VII. The Storage Tanks:

The three (3) component tanks are used as:

1. Condensate Storage tanks - these are tanks needed for storing condensate or light liquid of hydrocarbons

2. Produced Water Storage Compartment - these are tanks utilized for the necessary holding of treated produced water created by the FPSO processing facility

3. Crude Oil Tanks - they are the large storage tanks designed purposely for storing the produced crude oil onboard the FPSO

VIII. Offloading Arms:

1. This equipment is characterized for transferring the processed hydrocarbons to the shuttle tankers, tanker vessels or pipelines for safe transportation

IX. Centralized Control Rooms:

1. The FPSOs are equipped with unique centralized control room facilities required for monitoring the onboard systems and processes of the vessel's functionalities

Advantages and Disadvantages of FPSO

There is a need to consider the choice between FPSOs and other methods of production in the oil and gas industry that are dependent on multiple factors such as the characteristics of the offshore fields themselves and economic considerations, including the environmental regulations.

However, each production method when analyzed has its merits and demerits, and decisions should be typically aligned on a case-by-case basis for more rational conclusions and below are some of the advantages and disadvantages associated with the use of FPSOs:

Advantages of FPSOs

1. **Vessel's Rapid Deployment:** The merits of using FPSO is that it can be deployed more quickly when compared to the fixed platforms construction, thereby enabling faster commencement of oil and gas production.

2. **Declined Environmental Impact:** Using the FPSOs can help reduce the environmental impacts as they are built with a smaller footprint in comparison to the fixed platform or structures. Additionally, FPSOs can easily be disconnected and relocated when anticipating severe weather events occurring, this could help decrease the risks of causing severe damage to the environment.

3. **Versatility and Mobility:** By the structural design, FPSOs are mobile deploy drilling vessels allowing easy movement from one offshore location to different fields when needed. This provides oil and gas companies with the flexibility to explore and produce hydrocarbons from various oil and gas fields in deep-water or remote locations.

4. **Cost-Effective:** There are many scenarios where the use of an FPSO can be considered less expensive than structural fixed platforms. Typical examples include deepwater or remote locations where the construction of fixed structures could be highly expensive economically.

5. **Integrated Processing Facilities:** The design of FPSOs is equipped with integrated processing facilities onboard to facilitate multiple functions, enabling effective separation and treatment of oil, gas, and water to be carried out directly at the offshore production sites.

6. **Optimum Management of Reservoirs:** FPSOs are equipped to facilitate optimal reservoir management through adjustments to

the rates of production in a well-structured interventional manner as required.

Disadvantages of FPSO

- **Capital Intensive:** In most cases, the construction and deployment of FPSO involves high initial capital cost in the form of significant payment upfront, this can be seen as a disadvantage compared to some other production methods.
- **Limit in Processing Capability:** FPSOs may be limited in processing capacity when compared to other processing facilities or larger structured platforms. This could affect the overall oil and gas production capabilities of the organization.
- **Dependent on Tanker Vessels:** FPSOs rely heavily on shuttle tankers or pipelines for their offloading processes of produced hydrocarbons. To some extent, this adds to the complexity of the logistics and additional costs to the overall production system.
- **Dependent on Weather:** FPSOs' vulnerability to extreme weather conditions is high such that, operations can sometimes be affected during severe hurricanes or storms. These situations could have the potential to cause production shutdowns.
- **Limited Space:** Space constraints on FPSO vessels may serve as a challenge for accommodating some of the additional processing equipment when needed. This can also affect the expansion of production capabilities of the oil and gas company.
- **Environmental Risk Associates:** Even though FPSOs can be moved easily from one offshore site to the other during the anticipation of severe weather conditions, the process of disconnecting and reconnecting can pose significant risks to the environment and those challenges could be detrimental when unguided.

SECTION IV

Safe Systems Of Work And Risk Management Methods In Oil And Gas

On completion this part four, health and safety professionals, students and managers should derive knowledge that would enable them to demonstrate core competence in Safe Systems of Work (SSOW) and Risk Management Methodologies in oil and gas industry.

Additionally, the 4th part of this book covers typical understanding on both familiar and unfamiliar conditions in the oil and gas industry that would help occupational safety and health practitioners when utilized.

The benefits in this section IV include the following:

- **Safe Systems of Work (SSOW)**
- **Hazards Inherent in Oil and Gas**
- **Sources of PSM Information**
- **Risk Management Methodologies**
- **Evidence Documentation**

ELEMENT SIXTEEN

SAFE SYSTEMS OF WORK

Safe systems of work, also known as SSOW is a formal procedure that requires a systematic and thorough examination of workplace tasks with the sole aim of identifying all hazards.

Safe Systems of Work are needed where hazards relating to the job cannot be eliminated, indicating the existence of risks.

The Safety term **"Safe Systems of Work"** is beneficial when it is utilized as a well-defined approach purposely to identify all hazards at every stage of the work through careful examination.

In general, SSOW consists of multiple methodologies that companies can select the most suitable to fulfill their industrial needs when undertaking a high-risk job to ensure all hazards are identified, risk assessed and precautionary measures provided to mitigate them.

Examples of Safe Systems of work includes Permits-To-Work Systems (PTW), Hazard and Operability Studies (HAZOP), Qualitative and Quantitative Risk Analysis, Hazard Identification (HAZID) and so on.

The SSOW typically in Oil and Gas industry, the uses of **Failure Mode and Effect Analysis (FMEA)** is equally helpful to determine the "Cause and Effect" via prepare-plan knowledge about systems and their itemized such as machinery, instruments, or equipment.

The implementation of FMEA analysis depicts more of (Rational Assumption Theory with practical application) that tries to determine what will possibly cause a machine to fail or equipment part to fail and the effects on that plant as well as the operational activities of the entire process the motive identify and prevent before disaster strikes.

Even though companies would choose differently, one or multiples among the above-mentioned procedures and others as their Safe Systems of Work suitable for the undertaken.

Factors to consider and key elements when designing an SSOW should conform with the following fundamentals;

A• The Key Elements of SSOW

I• The SSOW should be formal

II• Specifically to serve as systematic approach used to examine work in order to identify the hazards

III• Defining Safe Methods to conduct tasks

IV• SSOW should be provided by the employer and made available at workplace

V• Only competent personnel should be assigned to deploy the SSOW

VI• Designers should consider the participation of employees when developing the SSOW

VII• Accurate documentation of SSOW is required

B• Factors to Consider - An SSOW being a set of well-defined processes according to which a task must be carried out at the workplace requires the importance to identify the job at hand and consider how the work should be performed.

The next is to determine the level of difficulties that might arise during the task execution with the potentials to expose you, or your employees to the risk and to ascertain these things, the following factors should be considered at the development stage of the SSOW

1• The People - determine who is for what? For example, a competent person assigned with a specified responsibility.

2• The Equipment - ascertain the type of equipment, plant, or machinery involved and decide if any other safety tool is required

3• Materials - Ensure to know what materials that might be used and their safe handling process during task execution, for example, chemicals and other hazardous substances people might be exposed to when performing the job

4• The Environment - Know the environmental conditions under which the work is being done, factors should include windstorm, thunder storm and lighting, temperature level, downpour (rainfall) and others.

It would answer whether or not if the task can take place, likewise Engineering or Technical control measures, procedural, behavioral or administrative control policies, PPE and so on.

Note: Adequate design of Safe Systems of Work should be integrated into the organization's existing Health and Safety Policy serving as improvements to achieve efficiency in managing and protecting the employees and others who might be affected.

It includes providing protection to the company's assets such as plant and equipment while safeguarding the environment at the same time.

The design of SSOW that aims to achieve effectiveness and efficiency in managing workplace hazards which cannot be eliminated from the job should follow the appropriate fundamental steps below:

i• Assessing the Job - Begin this part by knowing the company's operational activities and determine the specific task to which SSOW is being deployed.

Safe Systems of Work should be made to address issues of hazards that are paramount and not trivial, therefore a vigorous assessment is required to cover every aspect of that particular task, examined and documented in writing to prevent an oversight of uncaptured elements about the work.

Writing the task assessment at this stage could be achieved as;

1• Assess What is being used at the site: This should cover existing items utilized in the operations. For example, Machinery, Hazardous Substances, Plant and Equipment, Electrical Power Sources and so on

2• Identify Sources of Potential Errors: Here, all potential error sources should be assessed such as Equipment Failure, Possible Human Error, Incorrect Assumptions, Mixed Messages and other Foreseeable Errors

3• Assess the Task Location: Know where the job is being carried out, a specific part of the workplace. Assess the working environment and its existing protection, determine additional protection needs where necessary

4• Assess How the Work would be performed: Write down how the task to which safe system of work is being developed should be carried out. Include task frequency, training needs for personnel, and procedures required to be used.

Note: Assessing the job when developing SSOW should be conducted by Supervisory Personnel alongside inputs from the field employees typically those with detailed knowledge of the activity.

This would enhance the work system effectiveness, makes it more practical and safer, as well as clarifying any incorrect assumptions that supervisors might possess about work methods contrary to the reality.

Additionally, consulting with workers who might be directly or indirectly exposed to the risks at the site is a regulatory requirement, indicating compliance with the law.

ii• Hazard Identification and Risk Assessment (HIRA) - Identifying all hazards related to the job, and risk assessed their impact is another crucial part when developing Safe Systems of Work.

This point is a legal requirement imposed on employers to conduct comprehensive assessment of all risks that workers and others may be exposed to, in order to provide the necessary control measures to mitigate them.

This approach is to be utilized to ascertain anything about the job with harmful potential to cause injury, or ill-health to the people or damage to property such that, risk assessment should be conducted based on the likelihood and severity. Make note of.

1• Identifying Hazards: Look for possible causes of harm, it should cover anything about job that has the potential to cause harm, injury or damage

2• Evaluating the Risk: Conduct risk assessments on all hazards identified and determine the level of risks associated with each of them.

This includes listing the task elements at all levels, and risk assessed each element by assigning outcome and consequences to them, then proceed to determine the outcome of events using the safety term "Likelihood" to those answers, find out how likely is the harm occurring according to the existing controls measures.

To ascertain the consequences of that harm when occurred, describes the "Severity" which also answers, to what extent would the people be injured or suffers occupational ill health, or the degree of property damage.

Determine who might be harmed and how? Analyzing this part, a thorough consideration should give to all who might be affected by the task being undertaking such as children, pregnant women, physically challenged or people of determination, and young employees who might be vulnerable.

Put in place additional control measures adequate to safeguard all those who might be affected, ensure to mitigate the risk to minimal level, for example, As Low As Reasonably Practicable.

Note: During this stage, if what is being anticipated signifies high loss potential, then introduce and implementing formal hazards analysis such as Hazard and Operability Studies (HAZOP), or Failure Mode and Effect Analysis Studies (FMEA) as your technique would be appropriate.

Conversely, If the potential loss is considered lower, then simple approach like Job Safety Analysis (JSA) would suffice.

iii• Define Safe Methods: The process of developing suitable and sufficient safe systems of work require that Safe Methods should be defined.

Provide in detailing how the task to which the SSOW is being deployed should be conducted without deviation in order to achieve the purpose of the systems itself, if hazards and risks about the job cannot be eliminated, then adequate procedures to ensure a safe work method must be provided as the appropriate alternative.

Even though different job requires different safe systems depending on the nature of hazards involved and level of risks, instructions from managers, supervisors, safety professionals and other competent personnel requiring employees to perform a particular task under well-designed safety management systems would be ideal.

Allowing workers alone to come up with their own work method is contrary to SSOW, (because one of the real risks, is doing nothing).

Chosen system can be explained based on written form, or even through verbal communication and should conform with the organization's general guide on risk levels, and the Safe Systems of Work requirements.

A written or documented records would serve as a value added to the entity, for example, future references and a defined methods could be categorized under the following:

1• Very High-Risk Task - Permit to Work

2• High Risk Job - Written Safe System/PTW

3• Moderate Risk - Written Safe System

4• Low Risk Activity - Written Safe System

5• Very Low Risk Task - Verbal Instructions could be utilized, alongside a brief written safety rules for conducting such job.

Note: All SSOW that requires to be documented should be written clearly and simply for the understanding of workers involved carrying out the job, avoid technical words and present a summary information to cover every important aspect of the systems.

Among other things, safe systems of work template that are typically used should contain the following:

1• Detailing how safely to conduct the task set up and the authorizations required.

2• The required conditions to be met before the commencement of the job

3• Major steps about task and the anticipated hazards to be noticed

4• The designed and approved safe working methods to be comply with, and should include procedures to reach and return from the task area

5• Procedures to dismantle an equipment and dispose of waste (hazardous materials) upon task completion

iv• Implementing the SSOW - The main goal of every Safe Systems of Work can only be actualized when systematically implemented and properly maintained.

The implementation stage requires assigning competent personnel to perform the task at hand, and should include the visions for adequate supervision during task execution to ensure adherence to the safety systems being implemented and also to ensure that workers comply with the SSOW requirements at all times, they must be:

1• Trained Adequately - Ensure that employees are trained adequately about the job specifics and what is expected of them because workers who have obtained adequate training turned out to know how to carry out the task assigned in the process.

2• They exhibit competence performing the task safely

3• Knowledgeable about the systems and hazards with the safe systems aimed to eliminate or mitigate.

4• They understand the established communication channels and how to utilize them in connection to the job being done.

Having a well-defined communication to be included in developing the SSOW is essential to achieve effectiveness and efficiency.

V• Conduct Review - The system would require a period review to determine its effectiveness, knowing the strengths and weaknesses of the system would be appropriate to introduce and implement additional mitigation policies when needed.

Reviewing and revising the systems regularly would be appropriate to analyze and identify changes in conditions that would create accidents or incidents.

Note: Under the review, frequent monitoring of the SSOW is required to assess the proper functioning of the systems at all levels, while the regular monitoring can be used to ensure the system remains appropriate for the task execution.

Check for the employee's compliance level, it would help answer if the safe work methods are being followed as required because, checking the systems after an accident or incident has occurred is inappropriate.

It will only describe the reactive part of systems to the team, while the active or proactive monitoring that aims at prevention is failing resulting in the incident anticipated, and while monitoring for deficiencies, use the basic questions below:

1• Does the employees still finds the system workable or effective?

2• Are the stated laid-down procedures being carried out? Look for deviations

3• If yes to #2, Are these procedures still effective for the task designed for? Look for defects

4• Is there any occurrence of change that requires the systems to be revised or reviewed?

ELEMENT SEVENTEEN

HAZARDS INHERENT IN OIL AND GAS

The level of risk in oil and gas industry and other petrochemical companies are high due to the inherent nature of hazards associated with those activities in the global economies where hydrocarbon and hydrocarbon-under pressure are being processed.

There are common terms used to describe different hazard that can be identified by the level of classifications in petroleum industry, also called the oil and gas. These are the Flash Point, Vapour Density, and Vapour Pressure.

The Flash Point - This term flash-point is used to describe the lowest temperature possible at which a certain vapour is considered to be sufficient to provide off to 'flash' when any ignition source is applied.

Categorizing the flash point could be achieved in three (3) different classifications.

Categories of Flammability

Flammable categories	Flash point
Extremely flammable	Below 0°c
Highly flammable	0°c - 21°c
Flammable	22°c - 55°c

Limits of Flammability

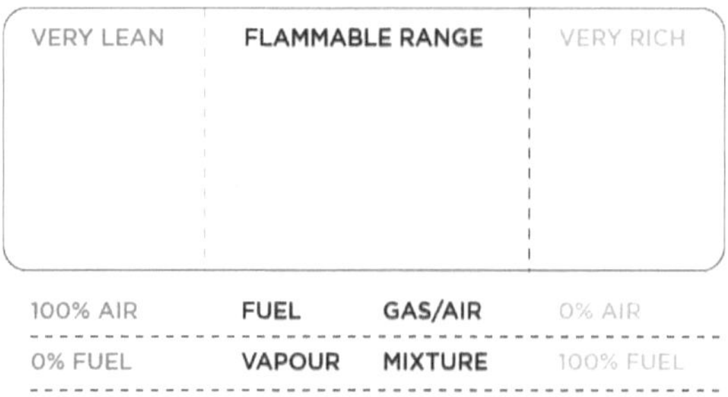

The Vapour Density – Vapour density basically defines the mass per unit volume of vapour.

The Vapour Pressure - This explains the level of pressure exerted by a vapour in an equilibrium value of its own liquid or other solid alternatives.

Causes of Explosives in the Oil and Gas

Evident proof used has suggested that, most of the explosions occurred in the process industries are significantly relative to the following circumstances as the causes:

I. Where metal tools have generated sparks

II. Sparks created by static electricity

III. During hot work, such as grinding, welding, metal cutting and others

IV. Where naked flames have occurred

IV. Sparks created by electrical equipment

TOXICITY AND ACUTE TOXICITY

Using the term Toxics in this PSM module, explains the degree to which the content of a poisonous substance or harmful chemical can cause harm to humans or animals whiles by extension, toxicity could see as the harmful effects on living organisms in general including plants in an environment when comes in contact with poison.

Acute Toxicity are dangerous and harmful to create immediate deficiency on living organism such as death, permanent disability or chronic sicknesses at single exposure or upon contact and must be handled adequately according to industrial standards till disposal without any limitations.

In addition, **acute toxicity** defines any harmful effects observed in an organism through a single contact of toxic substances or hazardous chemical compound short-term exposure

Furthermore, in the context of inherent nature of hazards in the oil and gas industry, the subject matter '**substance toxicity**' describes the Ability of any Chemical Molecule to cause Injury or Harm upon reaching a Susceptible site in the Body.

In this case, both toxicity and acute toxics can also refer to the application of the Quantitative Study of the human's body reacting to a toxic substances or poisonous chemical exposure and its effects such as long-term, short-term and among others.

The human toxicity is relative to the potential harm of a unit of chemical release into the environment in relation to hazardous materials at workplace in the process industries which is solely based on both the inherent toxicity of a compound and its potential dose.

The extent to which a toxic substance can affect the body could be analyzed based on the following three (3) basic entities representing the fundamental classifications, namely; Toxic Chemicals, Biological Toxics and Physical Toxics.

Chemical Toxicity: These are harmful materials with its toxic nature has the coverage of the Inorganic Substances such as Lead, Hydrofluoric Acid, and Chlorine Gases which are commonly associated with the pharmaceutical industrial Organizations.

Organic Toxicity: These are a compound substances of methanol gases, a toxic of colourless, volatile flammable liquid alcohol, prepared chiefly by oxidizing methane alongside hydrocarbons preparations from the petrochemical installations.

In generic terms, the coverage of chemical toxicity extended to other toxics such as polar organic maker compounds and toxic organic substances would have the following as a Tabular Representation of Chemical Compound Preparation (CCP's) and their Chemical Symbols (CS's), sometimes referred to as their Abbreviation Forms (AF's) or chemical formulars (CF's).

Examples of Toxic Preparations with Chemical Symbol Identifications

Chemical Compound Preparation	Abbreviations / Chemical Symbols
Polychlorinated Biphenyls	(PCBs),
Organochlorine Pesticides	(OCPs),
Polybrominated Biphenyl's Ethers	(PBDEs),
Poly-cyclic Aromatic Hydrocarbons	(PAHs),
Nitro- derivatives	(N-PAHs),
Dicarboxylic Acids	(DCAs).

Focusing on the inherent hazards in oil and gas industry and their risks, '**toxicity**' are poisonous substances when enters into the human body could damage an enzyme.

To some extent, they prevent the body function from Haemoglobin production in the blood, which could result in age accelerating, leading to the failure of energy production, and would lower the body's protection against oxidized stress.

The prolonged failure of these normal body functions has the potentials to increase the person's risk of diseases such as Cancer and others which may be untreatable causing early death.

Aiding the International Labour Organization ILO's aims that seeks to ensure workplace safety for all and good practices in managing the environment where industries operate while protecting the ecosystems, some of the best practice countries who are also pioneers in the Occupational Health and Safety management deserves to be acknowledged.

The ILO member states such as, Work Safe Australia (WSA)

United States Occupational Safety and Health Administration (OSHA), United Kingdom UK, Health and Safety Executive (HSE), Institute of Occupational Safety and Health (IOSH) and others have contributed immensely with different authorships for training HSE professionals worldwide.

Such progressive inputs helped promotes and improve the workers' health and safety, and other stakeholders for a safer environment, human economic development and communities' well-being.

The British Health and Safety Executive (HSE) have a well-designed knowledge with different authored publications to determine or measures the limit of workplace exposures to employees, titled "Workplace Exposure Limit" by HSE UK.

Below explains some toxic substances and their preparations:

1. Extremely Toxic - this defines any substance which contains higher level of toxicity (very harsh, malicious, or harmful) that needs to be handled using the right prescribed procedures at all times.

This is because even in very small quantities has the maximum potentials and can cause immediate death, acute or chronic damage to a person's health when exposed through the processes of inhalation, swallowing, or absorbed via the skin.

2. Toxic - these are poisonous substances, very harmful and unpleasant. In small quantities has the maximum potential and can cause death, acute or chronic damage to a people's health when inhaled, absorbed through the skin or swallowed.

3. Harmful - substances with a potential of causing or likely to cause harm. When exposed can cause death, acute or chronic damage to a human's health when inhaled, swallowed or absorbed through the skin contact.

4. Corrosive - this has the potential of destroying living tissues on contact or through body contact. An example could be Hydrofluoric Acid, a solution of hydrogen and fluoride (HF) in water. This solution example is colourless, acidic and highly corrosive.

5. Irritant - these preparations have the potential to cause inflammation or discomfort to the body. They are non-corrosive substances to which in an immediate contact, prolonged or repetitive with the skin, or mucous membrane, can cause inflammation.

Substances effects, if inhaled, ingested or penetrate the skin:

I. Sensitizing - A sensitizer substances could be responsible for either skin defects or respiratory in humans because, it has the potential or capacity of eliciting a reaction through hypersensitivity to the airway.

Long-term exposure to it characteristics can produce adverse effects resulting from the severe outcomes of being in contact repetitively affecting internal organs essential to support breathing while, the skin sensitizer on the other hand is a substance which has the potency to induce an allergic response when contacted with the skin physically.

Therefore, sensitizers' substance may be sensitizing by either through inhalation or physical skin contact and Raising Safety Awareness (RSA) for the use of these industrial chemical preparations is a necessity in forecasting Personal Safety (PS) within Process Safety Management Systems (PSMS) to attain complementary and the integration needs to promoting the welfare and general well-being of those who might be affected when unguided.

• **Carcinogenic** - These preparations or substances has the potential to cause cancer, induce or increase its incidence.

• **Mutagenic** - Mutagenic agents could be either chemical or physical that has the ability to modify the human genetic code in a harmful way and the change that may occur in the person's genetic code is referred to as Mutation.

Throughout the existence of an individual's life, people accumulate many mutations from varying sources into their cells, a mutagenic agent is that which may induce Heritable Genetic Defects (HGD's) or increase its incidence.

• **Toxic for Reproduction** - The general concept of reproduction refers to the process through which living organisms' increases in number. Naturally, most living things produces young ones of their own kind and so do humans.

Reproductive Toxicity - describes the adverse impairment on the sexual function and fertility in adult humans both male and females, include the adverse effects on development of the offspring.

This toxic for reproduction is substances of non-heritable adverse effects, instead it may cause or increase dysfunction of male and female reproductive functions or capacity.

Chemical Properties and Hazards of Gases:

Hydrogen (A Catalyst Re-generator, In Petroleum Refining)

• Highly Flammable or Explosive (Explosion Range from 4.9 - 75%)

• Colourless and Odourless nature

• Low Density

• Low Ignition Energy

• Non-Toxic, but Asphyxiant in high concentration

• Reaction, Vigorously with Oxidants

Methane (Natural Gas / NG)

- Highly Flammable or Explosion (Explosive Range from 5% - 15%)
- Low Density (Explosive mixture may be created below low-ceilings and so on)
- High Ignition Energy / Easily Ignited
- Simple Asphyxiate
- Odorous / Contains Odorising Agent

Liquefied Petroleum Gas / (LPG) Family- Propane or Butane - (Fuels)

- Colourless and odourless
- Highly Flammable
- Easily Liquefied Gas
- Reacts Explosively with Chlorine
- Denser than Air, collecting at low level
- Ignited Easily
- Simple Asphyxiate and prolonged inhalation may cause drowsiness
- Explosive mixture form, readily flashing back to a leak source liquefied

Natural Gas / (LNG)

- Highly Flammable or Explosive Gas
- Liquid gas vaporizes easily
- Easily Ignited

- Colourless and Odourless
- Volume increases 630 times during vaporization
- Simple Asphyxiate but non-toxic
- Contact with liquefied form will cause frostbite. (Very cold in nature boiling at 161°C)

- Frostbite, an injury to the body tissues caused by an exposure to extremely cold, usually to the finger, nose, and toes and persistent or prolonged exposure can produce adverse effects or death in worst case scenario.

Such substances are highly common with a specific fire extinguisher called Dry Chemical Powder abbreviated as DCP and must be handled with caution, do not let the outgoing gas from the portable extinguisher touch the naked body.

Nitrogen

- Colourless, Odourless, and Tasteless
- Non-flammable
- Used as 'Cover' layer of gas on top of flammable and explosive substances
- Nitrogen content in the blood reduces the amount of oxygen-carrying capacity in the blood
- Used to free pipes and pipeline purging
- Used for 'Inert' flammable and explosive atmospheres vessels

Hydrogen Sulphide

- Toxic, irritates eyes, skin and respiratory tract and may cause respiratory paralysis when exposure is persistent
- Flammable gas

- It has the capacity to travel long distance and flash back when spark is applied
- Colourless but has strong smelling similar to rotten eggs
- Denser than Air, gather at low levels

Oxygen

- Violent reaction with oil and greases
- Colourless and Odourless
- Non-flammable but can support combustion
- Oxygen Enhancement may cause fire and explosions.

Anti-foaming Agents and defoamers

The words Defoamer and Anti-foaming Agents are often used interchangeably to represent their functions to control foam created or prevention further generation of foam in the process industries.

Even though both agents are used for a common intended purpose, they can be distinguished slightly such that, the chemical term Defoamer, is used specifically for the removal of existing foams that has already been formed, and Anti-foaming Agents on the other hand are used purposely for the prevention of the formation of foams.

These are chemical additives that minimizes and hinder the creation of foams in industrial process liquids, they are used in the process and cooling liquids with the aim of decreasing problems arising from foam formation by dissolving or trapped air such as below:

1. Cavitation, that minimizing pump efficiency and also generating

noise

2. Decreasing the function of pumps and storage tanks
3. Control Bacteria growth in the fluids
4. Blocking of filtration equipment
5. Decreasing the effectiveness of the fluids in use
6. Mitigate Dirt and Debris creation and surface flotation
7. Shortened fluid enrichment times and increased costs
8. Prolonging cleaning and maintenance functions.

Anti-wetting Agents

These agents such as waterproof barrier coating is used to control problems cause by water or issues arising from wet sources.

- Anti-wetting agents provide protection against water ingress in harsh environment
- It also include some of the protection against corrosion
- Corrosive preventatives for fuel systems and process pipelines
- Other Refrigerants, for example, Propane, Ammonia, Sulphur dioxide and Methane

Micro-herbicides anti-bacterial treatments

I. Oil system biocides - found in oil production, and water injection systems

II. Water injection systems biocides

III. Fuel preservation biocides

IV. Water system biocides - typically, for salt and fresh water systems

V. Other Special biocides - these are special designs for example, to reduce sulphate, minimizing bacterial in drilling and process structural platforms and pipelines

Hazards and Risk Mitigation for Additives

HAZARDS

• Hazards at this point may not be necessarily categorized, rather it would depend solely on the Inherent nature of Chemical Hazard, Physical Form and the route or process of entry into the body.

RISK MITIGATIONS

- Hazardous Substances Risk Assessments / Analysis
- An Automated dosing is preferred, rather than hand-dosing
- Operate highest level of safe storage and handling procedure
- Ensure suitable Personal Protective Equipment PPE, examples are, Goggles, Respiratory Protective Equipment (RPE), Chemical Resistant Clothing, Breathing Apparatus and others.

WATER AND STEAM

The fundamental functions of water and steam in process are used as system cooling, lubrication or drilling mud's, fire deluge systems, and Advanced Hydrocarbon Recovery Procedures.

HAZARDS

- An injection of fluids into the body by pressure
- Severe steam burns may be sustained and (may include the lungs)

- Asphyxiation Control

MEASURES / SAFE HANDLING

- Design Special methods
- Ensure the uses of water and heat-proof clothing.

MERCAPTANS

These are group of sulphur containing chemical substances in the process industries, especially oil and gas.

HAZARDS

- May contain Offensive Odour (used in the process preparations as odorising agents)
- H2S and mercaptans eliminate in oil refineries and natural gas processing plants
- Effects such as Headaches, nausea, coughing, eyes and lungs irritation.
- At a very high concentration levels, causes difficulties in breathing, cyanosis, or (turning blue), loss of consciousness, as well as muscle spasms.

MEASURES FOR MITIGATION

- Ensure the wearing of Suitable Respiratory Protective Equipment or RPE especially when anticipated any potentially harmful levels.

Drilling mud or Drilling fluids

These are special preparations designed appropriately for the use in deep holes in oil and gas extraction.

- Lubricant (this minimizes friction and heat, and reduces the potency of friction-related complications during extraction processes)
- Carriers are strictly for the material through which drilling takes place
- Varying degrees of drilling mud for different circumstances (often relative to viscosity and density)
- Two most common types used are the
    1. Aqueous or (water) based,
    2. Non-aqueous (oil) based or Gaseous fluids
- It has a nature of Mineral or Synthetic

HAZARDS

- Contact with additives, (for example, Anti-foaming Agents, Fumes, Diesel oil and so on)
- Fire or Explosion, (for example, Flammable materials such as Natural Gases that can returned to the drilling work area) may cause explosion or increase fire incident

CONTROLS

- Provide fire safety precautions, for both proactive and reactive functions
- Provisions for Appropriate Personal Protective Equipment PPE.

Sludges also known as (drilling waste)

- This classification originated with an industrial term as "Low Specific Activity" abbreviated as (LSA) - sludges may contain some properties that are naturally occurring radioactive materials or (NORM).

Examples may include, Uranium, Thorium, Radium, and Strontium.

- Radiations contained by the above-mentioned NORM's are monitored through the setting-out areas
- The process of removing Low Specific Activity or LSA scale from manufacturing equipment uses (specialist dispersal chemical or high-pressure water flushing alternative)

Control Measures

- Provide Protection of Personnel typically from the exposures of body contact and inhalation.
- May required restricted, and controlled areas
- Allowable employees should be classified for high levels of radiation
- Phosphoric iron, requires special control measures to dispose of it.

In petroleum or oil and gas production, this Low Specific Activity / LSA scale can be found typically in the following areas:

1. Safety valves
2. The production wells
3. Inside separators / Water separators
4. Well heads
5. Production manifolds

ELEMENT EIGHTEEN

SOURCES OF PSM INFORMATION

This element describes all the important information and their sources required for the safer operations and maintenance of process plant, and systems include design, installation and safe transportation of hazardous materials.

Process Safety Management Information that are necessary should include:

1. Documented information that are,

2. Reliable, means accurate according to what must be done, by who and why, how, where and when,

3. Current, means it must be up to date and not sources information provided by and for about 10years ago,

4. Must be readily available commonly accessible to the people who need to use them. The information can be grouped into two parts:

Firstly, Information Internal to the Organization. Sources process Safety Management (PSM) Information that are considered internal basically referred to all the available information on the PSM within a particular organization.

The content is meant purposely for the use to safeguard all the procedures associated with the firm's operations in respect of controlling hazardous substances thereby ensuring compliance with the law.

Internal documents are often relevance to the oil and gas installation or process entity's potentials to cause harm or ill-health to people or property damage and environmental pollutions alertness that has been captured with the required mitigations.

Internal sources of PSM information differ from one company to the other due to the type of chemical they manufacture or work with, equipment or machinery they used, geographical location, entire plant size or installation facility and others.

The internal information is meant for the effective control of all the anticipated hazards and their risk associates to protect all stakeholders who may be affected by the release of these hazards to ensure they are safe.

Sources of PSM internal information are as follows:

i. Safety Data Sheets, abbreviated as (SDS),

ii. Process Design Criteria,

iii. Process Flow Diagram also known as (PFD),

iv. Standard / Safe Operating Procedures, or (SOP's),

v. Inspection, Audit and Investigation Reports,

vi. Maintenance Records,

vii. Piping and Instrumentation Diagrams (P&IDs),

viii. Process Control Systems,

ix. Relief System Design,

x. Fire Detection and Protection Plan etc.

The above mention source of PSM information may be considered general in the process industry while companies, have their own designed suitable to their needs therefore, those internal sources of information could be more than the number in listed samples given above.

Secondly, the External Sources of PSM Information. These are the multiple sources of information external to the organization and can be obtained through different channels worldwide and they serve as regulations and standardization of best practices supervised by authorities of different levels ranging from national and international levels.

Below are some of the PSM External Sources of Information listed:

- The EU-Directives, Parent Organization, European Union / European Commission
- The UK Health and Safety Executive (HSE), parent organization, Parliament of Great Britain
- The US Occupational Safety and Health Administration (OSHA), parent organization, United States Department of Labour
- Harmonized European Standards
- British Standards
- The International Labour Organization (ILO), parent organization, the United Nations (UN)
- Trade Associations and Professional Bodies, for example Institute of Chemical Engineers (IChemE), American Institute of Chemical Engineers Process Safety and so on.

ELEMENT NINETEEN

RISK MANAGEMENT METHODOLOGIES

The general risk assessment techniques and purposes are to identify potential hazards, understand the risks they pose analyse the risk for evaluation and decide on mitigation policies (most suitable) to be implemented.

Ensure to document your findings through appropriate recording procedure, indicate if additional controls are required also called (Safety Recommendations) then plan for review and update to determine effectiveness and efficiency.

Prevention is the common goal to achieve a safer work environment aiming at the protection of human lives, property damage, maintain the company's good public image (reputation) and avoid legal consequence should be considered vital to the process entities.

Risk management techniques used in the process safety management system or PSMS in the (process industries, PI's), specifically in the oil and gas sectors may vary depending on the organization's size, the operational activities, the level of risk the entity has to deal with, and the stakeholders involved in the risk management.

The Golden Rules States that, the higher level of risk, more depth and technical input would be needed and vice versa, especially for the successful management of more complex installations multiple risks associated with oil and gas processes.

The fundamentals methods for managing risks in the Oil and Gas companies can begins with five (5) steps approach which can be considered uniformity or standard.

Step 1. Identify the hazards - at this first stage, hazards identification defines the techniques used to determine or identify anything at the workplace that has the potential to cause harm or injury to people, or damage to property or both, for example, the workplace could be an oil refinery plant and the hazard is fire.

Step 2. Identify who might be harmed and how? - this method is needed and used to determine the answer about what the workplace hazard can do, who can be affected when it does, and how?

• Who can be affected or harmed means, an individual in person, for example the process engineer

• How can he or she be harmed, for example burn to death or sustaining 1st or 2nd degree burns from fire or explosion.

Step 3. Evaluate the risks and decided on control measures - the third part of these steps, is a strategy used to provide precautionary measure to control the risk.

• **Firstly of all,** you can evaluate the risk by gaining understanding about what you dealing with. Risk is determined by the likelihood and severity, or in other words (outcome and consequences).

• **Likelihood,** how soon can this oil leakage and hot work (welding/metal cutting) catches fire. The combination of sparks and fuel, when can it result into fire incident. Response could be eminent, or certain, very likely, likely, unlikely- means- it may happen, very unlikely.

• **Severity,** if fire incident occurs, may result in fatality loss-of-life death, major injury, over 3-days injury, injury / injuries treatable by first aid or (first aid injury), and minor injury.

Analyzing the risk is for the purpose of determining above mentioned in step 3, the answer would then be used to plan the mitigation policies as an adequate control measure.

Step 4. Record your findings and implement them.

During this stage, you are expected to write down everything obtained from the risk assessment and then proceed with the control measures to make sure no one is harmed, and all the company properties are safe.

Achieving this, would be enough justification that a great job has been performed by the risk management team and at this point, hazards control and all your mitigation policies should follow the prescribed Controls in the below diagram.

Hierarchy often starts from

NO.	Hierarchy of Controls / Control Hierarchy
1	Elimination - Remove hazards from the work / discontinue or substitute for non-hazardous
2	Engineering / Technical controls – Isolate the hazard / change how people work / substitution for less hazardous
3	Procedural or Administrative control - Take the people from the hazard; breathing apparatus, safety or isolation clothes etc
4	Personal Protective Equipment PPE - For example, Helmet, goggles, gloves, masks and so on.

Step 5. Review your assessment to make sure it is current and suitable for the organization.

This final stage five (5), requires the periodic revisits, a rechecked technique to determine effectiveness and efficiency because hazards can be unpredictable, especially in the PSM industries relation to hydrocarbon production processes.

These are installations where at every stage of the life-cycle of plant used by the company, new hazard would be introduced therefore previous control measures may be inadequate, for construction industries, every stage of work process has different hazards and require different risk assessment.

For example, construction works of building a new storage tank which begun from the foundation has now reached 23 feet high above the ground, such condition poses a higher risk when compared to the previous hazards identified at the beginning of the work at the ground level, you're dealing with fallen-from-height.

Risk management techniques may be applied differently by either statutory or regulator's requirements, the oil and gas company's policy, or based on international standardization.

In the United Kingdom (UK), all installations operating Offshore, has been regulated according to Safety Case Regulations 2005, which requires:

1. All hazards that have the potential to cause a major accident have been identified,

2. All major accident risks have been systematically evaluated,

3. And control measures have been, or will be taken to mitigate the major accident risks to ensure compliance with the prescribed law.

Under this condition, the PSM industries term used for the British law is known as **"Compliance Demonstration"**.

Hazard identification - this PSM term is considered as the CORE of comprehensive risk assessment such that, all hazards must be identified first, before risk estimation and ranking of risks can be conducted from highest level to the lowest level.

The next would be risk evaluation and implementation process, and then review. The main steps of hazard identification are as follows:

I. Hazard identification number one

II. Risk estimation and ranking of risks

III. Risk evaluation and implementation aim at risk reduction to ensure compliance with the law.

IV. Plan for Review.

Note: The first three (3) stages focus on Identification of Possible Additional Risk Reduction, despite their applications are different since they follow each other, the intended purposes are one and the same.

TYPES OF RISK ASSESSMENT

To avoid implications that comes with inaccuracies in risk assessments, techniques used has been divided into three (3) namely, Qualitative, Semi-qualitative, and Quantitative risk assessment methods, for classical explanation, use below table:

Risk Analysis Tools	Qualitative risk assessment technique (Q)	Semi-qualitative risk assessment method (SQ)	Quantitative or Quantified risk assessment technique (QRA)
NO.	Qualitative	Semi-qualitative	Quantitative
INTRO	The use of qualitative methods describes the risk assessment technique in which personnel is to determine frequency and severity of risks.	Using this technique means personnel needs to determine the risk frequency and severity based on approximation principle within a certain range.	Using this approach is where full risk quantification is demonstrated.
1	Use this particular method, if it is adequate for deciding on appropriate control measures.	Use the (SQ) method if it is adequate for deciding on appropriate mitigation controls.	If adequate, use quantified risk assessment QRA.
2	Record all the findings and recommendations.	Record all findings and recommendations.	If inadequate, increase the depth of the risk assessment model unit used, it would provide answers to all questions.

3	But if (Q) is unsuitable or inadequate, then use Semi-qualitative.	Again, if the (SQ) technique is not adequate, the first approach is to increase the depth of modelling of the risk assessment to ascertain if it is now matching the requirements.	Record all findings and recommendations.
4		If it does, record the findings and recommendations.	
5		If it does not, this means inadequate, then use (Quantified Risk Assessment or QRA) method.	

In the publication of British Health and Safety Executive, UK HSE, in their Offshore Information Sheet # 3 / 2006, they outlined more industrial specific guidance on how to utilize and also determine which among the above-mentioned risk assessment method is appropriate.

Samples of Starting Point Approach

NO.	Examples below could be used for Starting Point Approach
i	Big Integrated platforms or nodal platforms in the North Sea - would be likely to require Quantified Risk Assessment (QRA)
ii	Installations of less complex, and those with smaller workforce such as drilling installations, normally unattended installations abbreviated as (NUIs) Semi-qualitative risk assessment (SQ) could be ideal
iii	Installations and platforms where a defined standards are used or benchmarks for design and risk reduction-sufficient risk assessment method would be Qualitative (Q)

Risk Estimation and Ranking of Risks in PSM industries

To estimate risk and rank them, use a diagram below, it depicts the level of risk from determined with by the principles of likelihood and severity, it has simple calculations in which higher numbers means danger and requires comprehensive assessment technique to further reduce the risk to the acceptable level.

The following Likelihood and Severity tabular representation is a simple formula with clearer understanding, it represents three different meanings that are shown in colors of Green, Yellow and Red as the interpretation steps of the 5 by 5 metrics risk assessment method given as:

Risk Estimation and Risk Ranking 5 by 5 metrics

LIKELYHOOD → / SEVERITY ↓	VERY UNLIKELY	UNLIKELY MAY OCCUR	LIKELY	VERY LIKELY	CERTAIN
FATALITY	5	10	15	20	25
MAJOR INJURY	4	8	12	16	20
MORE THAN 3 DAY INJURY	3	6	9	12	15
FIRST AID INJURY	2	4	6	8	10
MINOR INJURY	1	2	3	4	5

The 5 by 5 metrics key

From 1-4 in green represent low risk	Maintain the existing controls
From 5-10 in Yellow stand for Medium risk	Additional control measures are required to reduce the risk further
From 12-25 in Red represent High risk	At this level of risk work must be stopped until further controls are introduced to reduce the risk

SYSTEMS APPROACH

The most efficient way to utilize risk management methodologies in oil and gas and process safety management (PSM) is for industries to understand the systems and their components in an organization, commonly known as systems approach.

To do this, begin with the heading,

<table>
<tr><td rowspan="4">RISK
MANAGEMEN =</td><td>1. Know your risks and what you should be doing about them as an occupational safety and health practitioner</td></tr>
<tr><td>2. Plan, prioritize, implement risk control measures</td></tr>
<tr><td>3. Ensure the effectiveness of your risk control mechanisms</td></tr>
<tr><td>4. Review and learn for continuous improvement</td></tr>
</table>

In order to achieve this, the United Kingdom UK, health and safety executive in the publication, the HS(G)65 SMS: introduced the following systematic approach.

Short Code	Policy - Organizing - Planning - Measuring - Auditing - Review
P	• **Policy** - this is the general information in details on how the organization's stakeholders who are responsible to manage risk should use, (the general statement of intent)
O	• **Organizing** – relates to the arranging of all the things needed to be put together for a function to work.
P	• **Planning** – is a specific design used as guidance on how to carry out a task based on its nature
M	• **Measuring** - the act of determining success or failure for further correctional improvements
A	• **Auditing** - this defines a design standard used to determine whether or not if criteria have been met.
R	• **Review** - this is to determine the level of performance by a function, or a task performed periodically.

Utilizing the Systems Approach

This forms part of the risk management techniques in the oil and gas or PSM industries which describes how to manage hazards and associated risks at each stage.

Managing hazards and mitigating risks at each stage using systems approach

INPUT ITEMS	PROCEDURE / PROCESS	OUTPUT / OUT-TURN
Design and construction	Routine and non-routine operations	Commodities or product and services design
Design and installation	Maintenance and repairs	Packaging and labelling
Purchasing, procurement and supply chain network	Plant or machinery and process modification	Logistics; storage and transportation
Selection or recruitment activities	Foreseeable and feasible emergencies	Offsite risks and onsite considerations
Selection and engagement of contractors	Decommission or shutdown	Disposal and hazards pollution control measures
Acquisition	Demolition or clearance exercise	Divestments or dispossession
Information		Information

Risk control of inputs section capturing (design, selection, installation etc)

NO.	RISK CONTROLS REQUIRED FOR:	Examples for application
I	PHYSICAL RESOURCES (Plant, Equipment, or Machinery)	Rigs or well, safety critical equipment, hazardous substances
II	HUMAN RESOURCES (Manpower Acquisition)	Selection or recruitment of personnel or staff and contractors
III	INFORMATION (From Internal and External Sources)	Health and Safety laws and standards Environmental Regulations/protection

Risk controls and hazards management utilizing process / procedures

NO.	AREA OF PROCESS RISKS	Examples for application
1	PRODUCTION WORKPLACE	Oil fields, Support systems, Facilities, Welfare, Access, First aid & First aiders
2	PLANT, MACHINERY OR EQUIPMENT AND SUBSTANCES	Oil or Natural gas drilling, Storage, pumping etc.
3	PROCEDURES	Shifts handover/takeover, Job design structures
4	PEOPLE	Leadership, Competence personnel

Other Vital Risk Management Tools

Below risk management tools are useful and recommendable for industrial safety in the oil and gas. They are terms in the industry that has been explained in this element.

I. HAZOP - (Hazard and Operability Studies). It is used for the following purpose,

- Purpose - to determine or identify any deviations from design intent, including their causes and consequences.
- Useful at product design stage of chemical and hazardous installations or processes
- HAZOP's function includes, multi-disciplinary team and brainstorming
- It uses guide words to identify or determine deviations, for example, LESS OR MORE
- It can be utilized to devise actions to reduce risk drastically or down to possible acceptable level.

II. HAZID - (Hazard Identification)

- Purpose - to identify all hazards at workplace for later risk assessment or analysis
- Utilizes "brainstorming", and often aided by key words
- Useful at the event of considering changes to already existing layout of plant.

III. Failure Modes and Effects Analysis (FMEA) - this risk management method is somehow purely a question???

- How can each component fail?
- Cause? What could be the possible cause

• Effects on system as a whole or partly? What would be the effects on the entire installation or partial with the affected components been equipment F, H, S, U&D

• How serious? What is the seriousness of the impact

• Can the failure be detected before effects become serious? This intends to seek for any or all possible mechanisms for prevention.

Conceptual Risk Management Methods

Industry Related Process Safety Standards

Increase the awareness of concepts of inherently safe and risk-based design. For example, **"Design It Out"**

Specimen to promote the concepts of inherently safe and risk-based design approach, commonly called Design It Out are shown below.

ITEMS	PRINCIPLES FOR DESIGN IT OUT	Examples
A	Hazard Elimination	Ensure to - Discontinue, Substitute or Non-hazardous
B	Likelihood / Outcome Reduction	Ensure to - Simplify, Clarify, Redundant systems and Removal of Ignition Source
C	Severity / Consequence Reduction	Ensure to - Substitution or Less Hazardous, Isolation or Separation, (reduced inventory / stock, spill containment plan & control

Sources of Written, and Recognized Good Practice of Information

• United Kingdom (UK) Health and Safety Executive (HSE) Guidance and ACoPs

• National or local government guidance - this could be a guidance from regulatory body at local, state or national levels

• International or National Standards for example (British Standard BS, CEN, CENELEC, ISO, IEC, and others)

• Industry specifics or secretarial guidance for example, (Trade Federations, Professional Institutions, and so on).

Hazard Realization Concept in oil and gas

This concept can be achieved when searching for answers to questions relating to the industrial risk management such as, What if?

For instance, Loss of containment causing Hydrocarbon Release. A tabular representation below are examples of this concept.

ITEMS	Factors to consider from Hydrocarbon Release	Examples
i	Major Sources / Reason	Piping systems, instrumentation, valves, flanges etc.
ii	Operational Causes	Incorrect operation, wrongly fitted equipment, (for example, gas compression) human error
iii	Procedural Causes	Non-compliance with procedure (human) crror

In line to utilize the general concept of **'Risk Management Methodologies'** this element was developed to cover multiple risk mitigation techniques as much as possible, the conclusion part is by considering the **worst-case scenario** options, while **taking it further.**

Considering the possibilities of working down on each scenario, begin to tackle the angle of eliminating the hazard or reducing some of the consequences and other anticipating probabilities aiming to create a more safer working environment.

Worst-case scenario

Worst-case Scenario	**What – Where – When - How**
What?	Major Hydrocarbon Release (25kg & above)
Where?	From instrumentation or piping, on the section of gas compression near to the accommodation unit. With an uncontrolled ignition source
When?	During highest occupancy and fire deluge system on manual operation mode
How?	Inappropriate or poorly maintenance, incorrect fittings and not tightened correctly.

Followed by

Taking it further

1	Hydrocarbon release was ignited by electrical fault in the facility
2	Fire and explosion engulf the accommodation's canteen
3	All the 40 workers having their launch in the canteen have been lost
4	The Gas compression have been destroyed by the blast and explosion
5	The lost 40 crew members & entire Gas processing unit (long downtime)

Followed by

*Examples of **If** and **Then** (Utilizing if and then to eliminate or reduce consequences and other probabilities under foreseeable possibilities)*

Use IF ....&......................... THEN

Relate to # 3 of Taking It Further If the entire 80 crew members were divided into four (4) and 20 in each groups scheduled and allowed them to be in canteen at a time knowing the proximity of the Gas Compression	**Relate to # 3 of Taking It Further** Then we might have safe 20 more personnel, reducing human lost from 40 to 20 means additional 25% of our team members would still be alive instead of 50% that left us
Relate to # 4 of Taking It Further If the accommodation canteen was NOT next or closer to the Gas Compression Plant…	**Relate to # 4 of Taking It Further** Then we may NOT lose nearly about 50% of our crew members, half of our entire team & Gas Compression unit
Relate to # 5 of Taking It Further If the fire deluge system was NOT on Manual Operation Mode, but in an Automatic Function …	**Relate to # 5 of Taking It Further** Then water will be available and discharge automatically to extinguish the fire to minimize the impact

Additional Risk Management Technical Inputs

Organizations, Safety Engineers, Managers, Supervisors and other stakeholders in the oil and gas sectors could also endeavour to use the following risk mitigation models:

- The **Concept of Risk Control Barrier Models,** also known as **"bow-tie"**

It should cover at least these basic elements, the initiating events and consequences. It must be backed by an ***Escape, Evacuation & Rescue (EER) plan***

Initiating Events Barriers Samples	***Consequences Barriers Samples***
1. Plant Layout	• Detection System
2. Construction Standards	• Emergency Shut Down (ESD)
3. Inspection	• Active Protection
4. Instrumentation	• Passive Protection

- **The Concept of Swiss-cheese Barrier Model**

Consider the use of Modelling Software for Risk Identification

This concept requires the use of both software technologies and human efforts in combination.

Software to monitor and detect the risks, and personnel to obtain periodic information about what the software has identified. The engagement of software uses could be as follows:

a. The evaporation rate of flammable liquids

b. The dispersion patterns of leaking gases or vapors

c. The types of likely effects and scale of any fire and explosions

d. The rate of pressure rises,

e. The maximum pressure intensity of thermal radiation, and

f. The blast zones.

Most risks associated with all of the above-mentioned can be identified using modelling software risk identifications.

ELEMENT TWENTY

EVIDENCE DOCUMENTATION

This element being the final section of part III of this book is design to elaborate on the systematic form on how an organization's evidence is documented.

Typical example would be UK's Safety Case Offshore, and Safety Report Onshore.

This could be a specific legal or regulatory requirements as well as good practice such that:

1. Notification needed to be sent to regulator at the design stage or an alternative case of when moved or change of use.

For example, Piper Alpha incident, the Offshore platform was initially design for only oil production, and change of use was later introduced when gas was added.

Eventually, it ended chaos, and killed more the one hundred and sixty people onboard

2. Regulators require entities to submit their safety case for offshore / safe report for onshore concerning each type of installation.

3. In addition, an installation is not allowed to operate unless Safety Case Offshore or Safety Report Onshore has been approved or accepted.

4. In this case, the regulatory body would conduct an inspection on the installation first.

5. At the core function of Documented Evidence, Installations or companies in the oil and gas industry are expected to have all their undertaking documents kept as evidence to serve below purpose:

• To guarantee that, duty holders design or operate their facilities safely according to regulatory requirements.

• To refer as measures in place to identify or determine the potential needs for mitigating major accidents or prevent them from occurring.

Specimen of the UK's Safety Case and Safety Report

The typical form of Safety Case for Offshore installations or platforms and Safety Report for Onshore use, are made up of the following content prescribed in the box below:

NO.	Identify major accident hazards - by using a prescribed risk assessment tool, like bow-tie diagrams, design calculations and so on	NO.	Evaluate major accident risks and measures taken or to be take.
I	Each hazard scenario	I	Identify each hazard or incident scenario
II	Threats to safety and their causes	II	Assess frequency criteria
III	Barriers to prevent those threats	III	Assess consequences criteria
IV	Consequences of each threat were to be realized	IV	Assess EER facilities and requirements

V	Recovery measures required	V	Identify and assess risk control measures against ALARP
VI	Factors that could escalate the hazard or its consequences	VI	As Low As Reasonably Practicable / ALARP

The final stage

- Ensure and make adequate preparations for audits and audit reports
- Consider adequate SMS in place
- Major accident prevention policies, specifically for the safety report and cases
- Determine or identify safety critical elements available in response to major accident hazards
- Provide details of the emergency plan for the facility.

The above provided details relating Safety Case and Safety Report are for the use in the United Kingdom was instituted by Health and Safety Executive formerly Health and Safety Commission.

SECTION V

Balancing Capacity, Safety And Efficiency In Hydrocarbons Storage

On successful completion of this section, safety professionals, engineers, learners, instructors and companies in the oil and gas industry would gain excellent knowledge in familiar and unfamiliar situations in managing hydrocarbons and their safe storage processes.

Individuals would also attain comprehensive understanding in the safer way to operate steam boiler and furnace in the process industry alongside the upstream and downstream management of oil and gas production.

Take home benefits under this section 5 covers the areas of:

- **Fire & Explosion Hazards Protection and Risk Management**
- **Safe Operations of Furnace, Boiler and Steam Management**
- **Hydrocarbons Safe Storage Management**
- **Upstream, Downstream and Midstream Activities**
- **Emergency Response Procedures**

ELEMENT TWENTY-ONE

FIRE & EXPLOSION HAZARDS PROTECTION AND RISK MANAGEMENT

FIRE AND EXPLOSION HAZARDS

Fire and explosions are one of the most common form of hazards associated with the process industries especially, petroleum production companies, due to the fuel nature in most of their products.

Fuels are the substance that generates heat. However, heat on the other hand is a confirmed element of fire, therefore to handle arrangement that produces a (Near Elements) of fire Fuel and Heat in the process industry requires the finest practices to control fire and explosions.

You are producing fuel using heat processes within oxygen environment, means you're dealing with the fire itself and must be managed accurately with no limitations at all times.

EXPLOSION

An explosion associated with oil and gas production processes defines the sudden release of energy alongside rapid expansion of gases leading to the creation of shock wave and heat with noise.

The common forms of this explosion can happen due to ignition from flammable vapor or gases with causes ranging from leakages or ruptures, including oil and gas industrial equipment failures allowing flammable substances to escape and when comes into contact with an ignition source, explosion occurs.

An appropriate management of explosion is paramount in the oil and gas industry as it could lead to a very devastating consequences such as fatalities, injuries, and damage to installations.

Preventive Measures

To prevent an explosion, oil and gas companies can utilize the following mechanisms:

1. Maintaining Equipment Integrity,
2. Design and Implementing Safety Protocols,
3. Conducting Adequate Risk Assessment,
4. Provide Appropriate Training for Personnel,
5. Introduce Emergency Shutdown Systems,
6. Frequent Equipment Inspection,
7. Provisions for suitable Firefighting Equipment,
8. Safety Measures like Fire, Smoke, and Heat Detection Systems,
9. Installations infrastructure designs focusing on minimizing the likelihood potentials of explosions, and
10. Adequate Gas Detection Systems etc.

FIRE HAZARDS

Fire in this case, defines any chemical reaction between Oxygen, Fuel, and Heat.

The oil and gas companies produce fuels under hot temperatures, and the involvement of high-pressure gases, including other activities with the potency that would lead to fire and explosion occurring.

Hot work activities such as welding, metal cutting or grinding a metal near Hydrocarbon Storage Tank must be controlled by permit-to-work (PTW).

The general view of fire and explosive incidents in the process industry requires a different scope to deal with, based on industrial specifics of fire and explosions which may differ from other forms of fires related to different entities.

For instance, using the same method of approach to handle fires in an Onshore head office of an Oil and Gas Company, say Ghana Oil Company (GOIL) would not be effective way to deal with its Offshore fire and explosion at drilling sites.

The same company's Offshore Oil platform which may be located on the Seabed and miles away from the coast must be managed without neglecting the **Basic Safety Function and Awareness** during fire emergencies.

Therefore, **define the term Fire as:**

F – Find

I – Inform

R – Restrict

E - Evacuate / Extinguish

Use F - Find, to identify the present of fire. For example, fire in the FPSO ship crew's accommodation

I - Inform, to communicate the presence or existence of fire to the people

R - Restrict, to condone the fire, try to control the rate of spread, and buy time by closing doors, widows etc.

E - Extinguish the fire, if reasonably practicable, (if safe to do so and does not endanger your health and safety)

E - Evacuate, conduct the evacuation, following the fire evacuation protocols, (Evacuate Personnel to Safety)

Note: The right spellings of Fire are FIRE remember to use them as mentioned avoid using FIREE as that is not fire, it simply implies that, during fire emergency all the first 3 steps are correct to be carried out, then decide which one among the last two is required. See example A and B below as your step-by-step approach.

Specimen A: Step • Find – Step • Inform – Step • Restrict – Step • Extinguish

Specimen B: Step • Find – Step • Inform – Step • Restrict – Step • Evacuate

Ensure to act appropriately, fire can double in size within a minute and your decision must conform with the fire safety parameters, for example, Time, Distance, Fire type and Size, number of personnel in the facility and so on.

Despite how well to utilize the above fundamental technique to handle fires, prevention is the most efficient way to deal with fire and explosion risks, especially in the oil and gas industry.

Below is the specimen of the fire triangle in the oil and gas industry, it contains the fire elements which forms the function of the triangle that must not be allowed to come together to create the triangle, because once the triangle is formed, fire would be experienced, finding the efficient way to manage the element of fire will help to prevent combustion.

The Fire Triangle

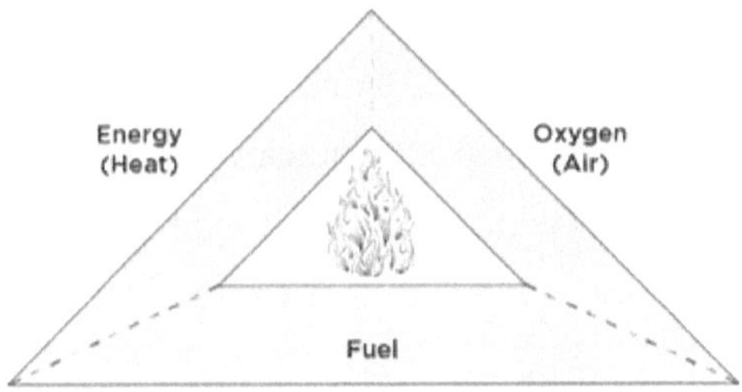

The fire triangle potential consequences:

- Explosion
- Thermal radiation
- Shock wave
- Fires etc.

Bring fire under control by removing one element (Oxygen, Fuel or Energy) From the Fire Triangle.

Textual Representation of Fire Triangle

In the oil and gas industry, fire is a complex thing and its chemical reaction involved the rapid combination of oxygen accompanied by fuel with heat release and the basic elements in every fire (oxygen, fuel and heat) are termed as the "Fire Triangle" and "Fire Tetrahedron" is what describes the chain reaction.

Elements of Fire:

I. Fuel: The Fuel part of fire defines any substance that generate heat and capable to undergo combination. Fuel is of many types and it can be solid, gas or liquid with common examples of gasoline, natural gas, paper, and wood.

II. Oxygen: For fire to exist, it requires oxygen from the air to facilitate the combustion process. When oxygen is present, it provides support to the chemical reaction needed to enable the fuel to burn speedily.

III. Heat / Light: Fire ignition occurs when a certain amount of heat is present to initiate the combination process. The origin of this heat is sometimes referred to as the ignition point or ignition temperature. Once the ignition temperature has caused fire to ignite, the chemical reaction would then facilities the fire to create its own heat using a self-sustaining process from the burning.

IV. Chemical Chain Reaction: Once fire has been sustained, an additional fourth element may be included known as the chemical chain reaction. This CCR is generated due to the continuous burning process resulting in breaking of some forming chemical bonds within the burning material that releases some level of energy in the form of heat, or light.

To manage fire requires thorough understanding and controlling of these initial elements for both prevention methods and firefighting processes. Once any of these three (3) components is removed, the fire triangle cannot stand anymore and fire can be extinguished.

Remember, the concept of the fire triangle would become incomplete if any one among the 3 elements is altered or disrupted.

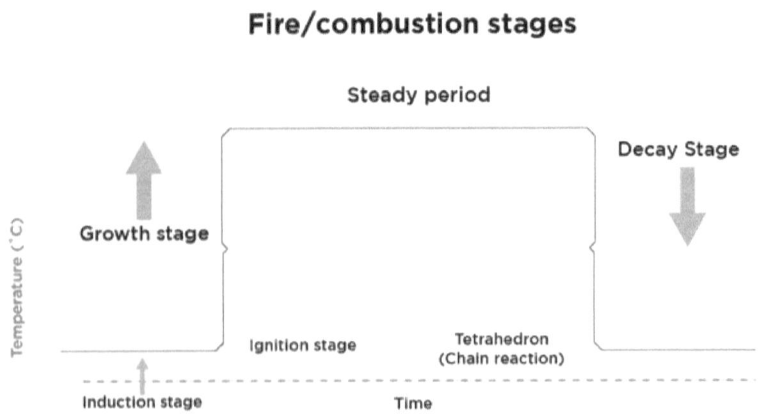

The diagram above, describes fire or combustion stages beginning from the stage of induction, to the ignition point, to the growth part, steady stage, decaying part and through to the tetrahedron.

Note: It is important to understand that, any good practices of fire safety must be designed with the main aim of manipulating and manage these elements to help prevent uncontrollable wildfires and to reduce the risk of fire harming the people and property damage.

Types of common fires in oil and gas

There are multiple kind of fires associated with oil and gas operational activities and they are commonly identified by an alphabet in English as an international standard with some designs containing both textual and diagrams showing what the type of fire entails.

Below is a diagram representation of fire classification table starting from Class A, B, C, D, F and K which can be mitigated in an oil and gas facilities.

Classification of Fire

Fire Classification	Class Explanation & Extinguishing Agent
CLASS **A**	**Class "A"** Fires defines primarily organic solid materials, also known as ordinary combustibles such as (wood, paper, fabrics, plastics) and so on. **Extinguishing Agent:** Water-based Solutions or Water.
CLASS **B**	**Class 'B'** Fires consists of flammable liquids like (petrol, gasoline, paint or solvent etc). **Extinguishing Agent:** Foam, Carbon dioxide or Dry Chemical Agents.
CLASS **C**	**Class 'C'** Fires involving flammable gases, examples include, butane, propane etc). **Extinguishing Agent:** Dry Chemical Powder **(DCP)** or Non-conductive Agents.
CLASS **D**	**Class 'D'** Fires that contains combustible metals such as (aluminium, sodium, magnesium, lithium) and others. **Extinguishing Agent:** Specialized Dry Chemical Agent designed for metallic fires.
CLASS **F**	**Class 'F'** Fires involving commercial deep fats or oil fryers. **Extinguishing Agent:** Dry Chemical Powder **(DCP)** or Non-conductive Agents.
CLASS **K**	**Class 'K'** These are fuel fires mainly from kitchen cooking oils, fats etc. **Extinguishing Agent:** Dry Chemical Powder **(DCP)** or Non-conductive Agents.

The illustration above outlines some common forms of fires in the oil and gas industry, ranging from fire classifications, brief explanations required for clearer understanding with their example materials when fighting them at their occurrences and the most suitable extinguishing agent that could be used.

ELEMENT TWENTY-TWO

SAFE OPERATIONS OF FURNACE, BOILER AND STEAM MANAGEMENT

Introduction

In general, furnace and boiler operations and steam management form an integral part of oil and gas production facilities because their functioning plays a major role in various processes such as power generation, heating, and other applications in the industry.

An appropriate boiler operations and steam management are essential to ensure more reliable and safe operation of oil and gas installations.

Activities such as frequent maintenance practices, adherence to the prescribed safety measures, and efficient control of systems can have significant positive contributions to the organization's overall production capabilities.

To manage the furnace and boilers successfully, below is the general overview outlining the uses of these industrial applications in oil and gas installations.

Furnace Operations

Furnace operations form an integral part of the oil and gas industry as it helps facilitate crude oil and natural gas production processes and the following points outlines some of the fundamentals in furnace operational activities.

Combustion Process - The functions of furnace cover how it combusts fuel to produce heat. The sources of these fuels can originates from oil, coal or natural gas, it could also be other hydrocarbons.

Regulates Temperature - Through it operational activities, furnace maintain the required temperature by making adjustments to its fuel supply and thereby controlling the combustion process.

Exchange of Heat - The amount of heat produced within the combustion chamber is exchanged another fluid or air that circulates through the furnace during operation.

Safety Mechanisms - Furnace are generally equipped with safety features such as pressure switches and flame sensors, as well as limit controls mechanism for ensuring safe and reliable operation.

Distribution Channels - Furnace distributes fluid or heated air to multiple parts of the oil and gas facility through a distribution networks like pipes or ducts.

Air Filtration Systems - Furnace are equipped with air filtration systems enabling the removal of impurities and particles from the air in circulation. This feature helps improving the indoor air quality, thereby preventing damage to the furnace components or parts during operations.

Required Periodic Maintenance - Periodically, maintenance activities are needed which includes cleaning, inspection, and regular maintenance. This is essential to ensure that, the furnace operates under maximum efficiency and safely.

By extension, this periodic maintenance should cover checking burners, replacing filters, inspecting heat exchangers and so on.

In conclusion, furnace and boilers plays a major role in the oil and gas industry and their proper operation and maintenance are essential requirements to ensure safety and maximum efficiency alongside the longevity of these systems in the production processes of crude oil and natural gas or hydrocarbons.

Boiler Operations

Firstly, boilers are used in oil and gas installations purposely to facilitate the generation of steam which in turn can be utilized for industrial processes including heating, power generation, etc.

Secondly, fuel boilers utilized in the oil and gas industry are commonly fueled using substances like oil, natural gas, and other hydrocarbon preparations depending on the cost and the availability of the selected fuel.

Thirdly, the fuel combustion processes that occurred inside the boiler produced high-temperature gases, which then transferred heat to the water and thereby generated steam and that steam is used for multiple applications in the oil and gas industry for the smooth running of operations.

Types of Boilers

There are two (2) main types of boilers, namely, fire-tube boilers and water-tube boilers.

1. **Fire-Tube Boilers:** These are types of boilers in which water surrounds the tubes through which the hot combustion gases pass through. These boiler kinds are commonly used for low-pressure and medium-pressure applications in the oil and gas production processes.

The fire-tube boiler sample

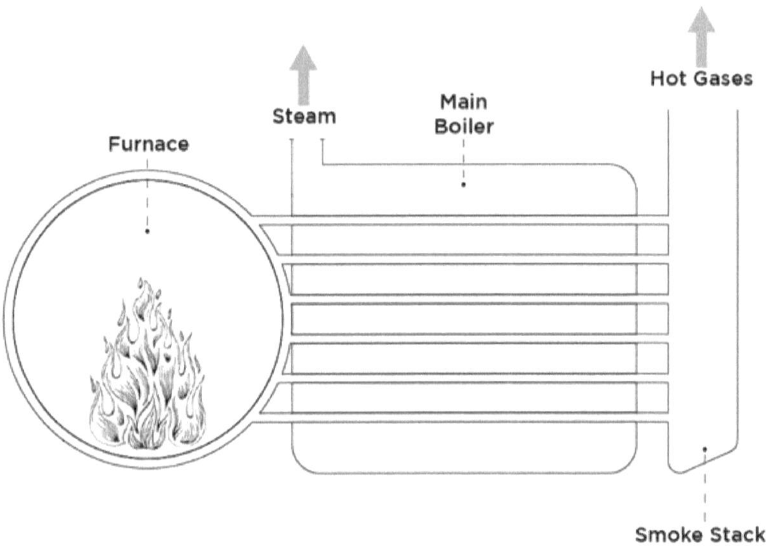

2. Water-Tube Boilers: With this kind of boiler, water circulates inside the tubes that are exposed to heat-generated sources or hot sources. These types of boilers are built with the suitability for handling high-pressure applications. Often, they are used for power generation in the oil and gas facilities.

The Water-tube Boiler Sample

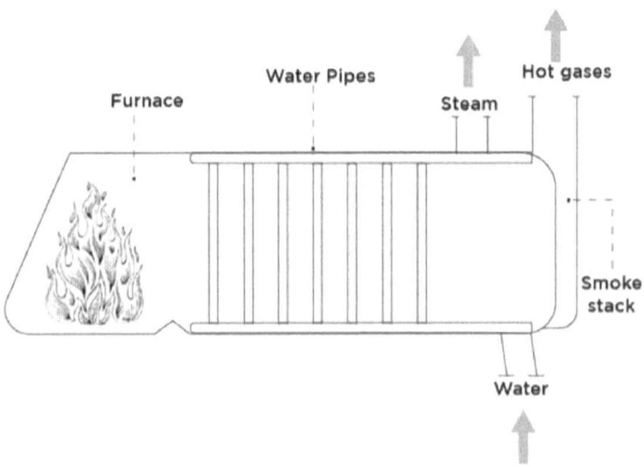

The Steam

Boiler operations in the oil and gas industry cannot be separated from the steam generation process since the steam management itself is tied directly to the boiler functionality.

However, discussing the boiler operations in the oil and gas industry, let us do a little dive into steam management, steam distribution, and safety considerations that concern the entire boiler activity.

Steam Management

Steam Management should cover the areas of steam generation, pressure controls, and temperature control.

i. Steam Generation: - Boilers through their functions create steam by heating water to reach its boiling point. Then the steam that was created is transported into different parts of the installations through pipes.

ii. Pressure Controls: - By the requirements of various processes, steam pressure must be controlled to meet them. Here, pressure management is maintained through the adjustments of fuel input. Fuel inputs are used to control the flow of steam. It can also be achieved by using pressure-relief valves.

iii. Temperature Control: - Steam temperatures are controlled for specific critical applications. In this case, control systems are regulated to influence the fuel inputs and feedwater flow to continue to maintain the required steam temperature.

Steam Distribution

This explains the piping systems and steam trap conditions in the process industry.

i. Piping Systems: - These represent well-designed piping systems used for the distribution of steam to multiple areas of the oil and gas installation. Insulation is commonly used to prevent heat loss in the transportation process.

ii. Steam Traps: - These are installed mainly for the removal of condensate or water-based from the steam when that particular steam cools.

Safety Considerations

Safety considerations should cover pressure relief valves and water level controls.

i. Pressure Relief Valves: - Boilers are designed with pressure relief valves necessary to regulate and control overpressures thereby ensuring the system's safety.

ii. Water Level Control: - Ensuring adequate water level in the boiler remains crucial to guarantee a safer and more efficient way of operation. Inappropriate or low levels of water can cause overheating and could lead to damage to the boiler.

Maintenance and Inspection: Conducting frequent maintenance, testing, and inspections are essential parts of ensuring safer and more efficient boiler operations and additionally, checking for leakages, cleaning boiler tubes, and inspecting pressure relief valves could help attain efficiency.

ELEMENT TWENTY-THREE

HYDROCARBONS SAFE STORAGE MANAGEMENT

Introduction

Hydrocarbons safe storage management is another vital component in the oil and gas industry to prevent environmental damage, oil spill, incidents; like fire accident and the release of poisonous substances.

Hydrocarbons such as, petrol, natural oil and gasoline, natural gas and others extracted from crude oil are flammable and capable to pose significant risks when they are not stored safely and handled properly.

Hydrocarbons safe storage is a multifaced procedure which includes the combination of technology, maintenance, engineering and procedural controls for the appropriate management of these petroleum products.

Steps to Store Hydrocarbons Safely:

1. Design and Construction of Storage Facilities: Construction and design of facilities to store hydrocarbons should be made as;

i. Facilities and storage tanks should be designed and constructed according to industrial standards and not limited to regulations.

ii. Tanks must be designed to contain compatible materials adequately for hydrocarbons storage.

iii. The design of storage tanks must be construted with materials resistant to corrosion and leaks.

2. Systems to Monitor Leaks Detection: Oil and Gas industrial organizations should;

i. Install leak detection devices for monitoring systems to immediately identify leakages or abnormalities that may occur to storage tanks.

ii. Install an automated system purposely to provide real-time data of hydrocarbon storage facilities with a well-programmed alarm for triggering and initiating responses in emergency situation where necessary.

3. Secondary Containment: Companies operating in oil and gas industry should;

i. Design and implement secondary containment mechanisms to remedy any leaks or spills that may happen, and this could serve as preventive measures of hydrocarbons from reaching the environment.

ii. The implementation of secondary containment should include berms, double-walled tanks, dikes and so on depending on the actual size and type of the storage facility.

4. Measures for Fire Protection: High considerations should be given to fire protection systems and must cover;

i. The installation of fire protection systems such as fire extinguishers, sprinklers, and foam suppression systems that are suitable for mitigating risks of fires in the storage areas of the oil and gas facility.

ii. Ensure to develop and implement appropriate fire response plans where necessary at all times.

iii. Design fire and smoke drills, conduct regular fire and smoke drills test as an effective preplan procedure for emergency situations when they occur, use the testing to know your emergency response preparedness during fire emergency.

5. Conduct Regular Maintenance and Inspection: To ensure hydrocarbons are stored safely, enterprises should;

i. Conduct frequent inspections on their storage tanks with the main aim of identifying and addressing any potential issues before they escalate.

ii. Develop and implement maintenance programs that are robustful to ensure the integrity of hydrocarbon storage structures such as tanks, valves and seals are intact for safe use, including storage transportation vessels and tanker-trucks.

6. Implementing Security Measures: To store hydrocarbons safely, adequate security would be;

i. Design and implement security awareness systems such that only authorized personnel would be permitted to access storage facilities.

ii. Implement additional security measures beyond verbal warning in the form of strict access control, CCTV cameras for surveillance, access keys registering, and others to prevent unauthorized access to the storage facilities.

iii. Ensure to protect all storage areas from intentional damage or vandalism act to maintain safety at all times, (consider effective management of both insider and outsider threats and prevent them).

7. Protection of Groundwater: This is the act of ensuring that;

i. Storage facilities in the oil and gas installations are located far from other sensitive areas like water sources and protected zones within the environment.

ii. Implementation measures are put in place aiming to protect groundwater from being contaminated in case of oil leaks or spill.

8. Vapor Recovery and Ventilation Systems: Organizations operating in the oil and gas industry can;

i. Develop and implement suitable ventilation systems to effectively prevent the buildup of flammable vapors within the storage areas.

ii. Develop and implement vapor recovery systems to effectively capture and manage hydrocarbon emissions, this would help minimize environmental impact.

9. Compliance with Regulations: Companies in the oil and gas sectors can endeavor to;

i. Stay adherence with local, national, regional and international laws that governs the storage of hydrocarbons as good practices to ensure safe storage management.

ii. Stay up-to-date by revising their practices on hydrocarbons storage to ensure they are currently aligning with industrial standards and as well as legal requirements.

10. Emergency Response Plan: It is required by the law that process industrial organizations should have (ERPs) for covering their hydrocarbon storage facilities and such could be achieved by;

i. Develop, implement, monitor and regularly review emergency response plans specific to ensure hydrocarbons are stored safely at all storage sites.

ii. Ensuring personnel are trained adequately on emergency procedures and can response quickly and effectively in case of oil spill, leaks or other incidents.

iii. Determine the efficiency of all training methods used, conduct drills and test to ensure accurate emergency preparedness are in place at hydrocarbon storage areas.

By adhering to the best industrial practices, and the content of this element, organizations can mitigate risks to acceptable levels, thereby ensuring that, there is safe storage of hydrocarbons in their oil and gas production processes.

ELEMENT TWENTY-FOUR

UPSTREAM, DOWNSTREAM AND MIDSTREAM ACTIVITIES

Introduction

The "upstream" and "downstream" are used to describe different production stages in the oil and gas industry, including the "midstream".

Upstream and Downstream are used to explain in detail of all the hydrocarbon processing procedures, from exploration to distributions and then to the marketing and sales, while the transportation and storage being classified under the midstream activities.

Utilizing these terms, upstream and downstream help to delineate the various activities occurring in the oil and gas industry before crude oil (hydrocarbons) can be accessed and consumed as petroleum products in our communities.

Upstream

In the oil and gas industry, upstream activities can be defined as the drilling, exploration and production processes of crude oil and natural gas and in this context, upstream operations focus on the oil and gas activities occurring in a place or near the geological source of hydrocarbons.

Key Activities of Upstream:

1. **Exploration:** This point describes the search for potential oil and gas reservoirs utilizing a geological surveying process, seismic studies and exploration drilling activities.

2. **Drilling:** This defines the process of extracting crude oil and natural gas from wells utilizing drilling rigs and FPSOs.

3. **Production:** This is the act of bringing hydrocarbons to the surface from underwater, and separating oil, gas, and water, as well as preparing them for safe transportation.

Downstream

Downstream activities in the oil and gas industry include the processing, refining, and distribution of crude oil and natural gas commodities.

In this case, downstream activities focus on transforming hydrocarbons in its raw state into usable end-products for consumers, examples of such usable commodities include petrol, diesel, natural gas, gasoline, jet fuel for aviation consumption and others.

Key Activities of Downstream:

1. **Refining:** This part defines the process of converting crude oil from its raw state into a refined products like petrol, gasoline, LPG, kerosene, jet fuel and so on.

2. **Distribution:** This describes the process of transportation and storage of refined commodities using tankers, pipelines, and other distribution networks to ensure the product reach end-users, (consumers)

3. **Marketing and Retail:** Here, the marketing and retail sector,

describe the sales of refined products and other petrochemical commodities to consumers through oil and gas stations fuel-stations, filling-stations, petrol-pumps and other retail outlets.

Midstream

This sector, midstream, serves as an intermediary stage between upstream and downstream of the oil and gas production processes and midstream activities cover the transportation, storage, and wholesale marketing process of crude oil, natural gas and other refined commodities.

Major Components of Midstream:

1. **Transportation:** This is the used of tankers, pipelines, and railroads to transport hydrocarbons between the production location, refineries, and distribution centers.

2. **Storage:** These involves the use of terminals and tanks to provide storage facilities for crude oil, natural gas and refined petroleum products to ensure the balance between supply and demand are met.

3. **Logistics:** Midstream sectors are companies that manages the logistics and other supply chain networks to ensure a steady flow of hydrocarbons in our societies at all times regardless of the community's geographical local, (every society needs fuel as energy for multi-purpose).

Gaining thorough understanding on the distinctions of upstream, downstream and midstream operations are pivotal for the stakeholders in the oil and gas industry.

Each sector of these 3 distinctions requires technical expertise, technologies, specialized skills and infrastructures to facilitates their smooth operations safely.

Upstream, Downstream and Midstream activities are required jointly to make up the entire value chain of the oil and gas industrial processes, beginning from exploration to end-user consumption in the global markets.

ELEMENT TWENTY-FIVE

EMERGENCY RESPONSE PROCEDURES

In the oil and gas industry, emergency response procedures (ERP) are comprehensive arrangements designed to address and manage unexpected incidents such as fires, explosions, leaks, oil spill and so on that may occur.

These incidents when they happen, poses significant threats to personnel, the environment and the integrity of the oil and gas installations and therefore, they are required to be managed accordingly to minimize the magnitude of impact associated with them.

Having a well-designed preplan arrangements to combat emergencies by ensuring the availability of a swift and coordinated response is one way to protect the environment, the integrity of facilities and people (employees, suppliers, contractors, vendors and members of the public).

When designing a suitable systems to address and manage unexpected incidents, oil and gas industrial organizations should consider the following major factors and include them in their emergency response procedures (ERP).

I. Planning Emergency Preparedness: This is the first stage of any good ERP and should include;

1. Develop emergency response plan in details, ensure to tailored

it to match a specific oil and gas facility's operations.

2. At this planning stage, considerations should be given to

- the nature of the activities,
- potential hazards, and
- regulatory requirements.

3. Established emergency response teams, allocate roles and responsibilities, and assigned them as their specified duties, (the team's specific duties), for example fire response team, response team for oil spill or leaks, emergency response team etc.

4. Ensure that, team members and personnel have been trained adequately to handle their tasks or responsibility assigned.

5. Ensure that team members and personnel are familiar with the emergency response procedures and understand their roles during emergency, different teams have different roles.

II. Risk Assessment: Companies operating the oil and gas sectors are required to include risk assessment when developing their ERPs and should conforms with;

1. Conducting thorough risk assessment in the oil and gas facility,

2. Identifying all potential emergency scenarios and hazards in connection with oil and gas operations,

3. Utilizing the outcomes of risk assessments to prioritize emergency response efforts, and

4. Use risk assessments results to allocate resources and implement the needed preventive mechanisms for oil and gas installation.

III. Emergency Shutdown Systems: These should include the following;

1. Install and maintain the required emergency shutdown (ESD) systems,

2. Ensure emergency shutdown systems are equipped with and can rapidly isolate critical equipment to prevent escalation of incidents,

3. Ensure to conduct regular testing of ESD systems to determine they are functioning properly.

IV. Emergency Communications:

1. Design and establish appropriate emergency communication systems in the oil and gas installations.

2. Ensure that systems are equipped with and can quickly and effectively notify personnel, emergency teams or emergency responders, and other relevant authorities when emergency occurred.

3. Ensure to implement emergency communication protocols required for oil and gas facilities, including area designated for emergency contacts.

4. Established a specified communication channels that are suitable for handling multiple or different kinds of emergencies.

V. Firefighting and Hazard Control:

1. Make provisions for both adequate portable and fixed firefighting equipment,

2. Conduct appropriate training for firefighters and personnel designated for responding to fire emergencies and other hazardous situations,

3. Ensure to implement control measures to efficiently mitigate oil leaks, oil spills and other potential environmental hazards.

VI. Safe Evaluation Procedures:

1. Design and establish safe evacuation protocols for the oil and gas facility,

2. Develop specific evacuation plans such as routes, assembly points for evacuees, and procedures for personnel evacuation when there is fire, explosion or other emergencies,

3. Regularly test evacuation plans by conducting frequent drills to ensure that all trained personnel or emergency responders are familiar with the designed evacuation procedures and assembly point locations.

Note: Fire Marshalls to conduct head-count of personnel in the installation after safe evacuation at the assembly point; Do this by calling names for responses, non-response personnel may be trapped and needs help to get to safety or safe area.

VII. Medical Response Facility and Resources:

1. Develop and establish medical emergency protocols such as first aid facilities to enable evacuation of injured personnel access immediate medical assistance,

2. Ensure that medical facilities in the oil and gas installations are equipped with first aid consumables and equipment peculiar to the facility's needs,

3. Ensure there are first aiders who are adequately trained as qualified personnel and available in the facility and they are well-prepared to respond to injuries and illnesses.

VIII. Coordination with Relevant Authorities:

1. Design, establish and implement communication and coordination channels suitable for reaching local emergency services, regulatory bodies or agencies and other relevant authorities.

2. Utilize collaborative efforts from external responders, ensure proper coordination and efficient response to address and manage emergencies.

Note: External responders may be needed to airlift personnel from offshore oil and gas platform engulfed in flames with wild spreading of fire making the airlifting the only viable means for rescuing personnel onboard the shift crew members

IX. Test ERPs for Effectiveness:

1. Ensure to conduct frequent emergency drills and regular training sessions to enabling personnel familiarize with emergency response procedures, use this to assess the effectiveness of ERP plans,

2. Utilize regular drills as ERP benchmarking to identify area for improvement, and

3. Strengthen the response systems weakness and update response procedures accordingly, ensure to put them in writing when updating.

X. Post-Incident Analysis and ERP Systems Improvement:

1. After an incident or emergency have occurred, conduct post-incident analysis thoroughly to identify both immediate and root causes and address them to assess the effectiveness of the ERP.

2. Use the findings of post-incident analysis to update and improve ERP preventive mechanisms and training programs, treat it as a need, as soon as possible.

SECTION VI

The Case Study, Environmental Impacts And Tasks Assessment

Upon successful completion of the last section, individuals would gain detail understanding on past incidents that have resulted loss of lives, property damage, and a huge financial loss due to the magnitude of the impact caused by fire and explosives (blown out), oil spill, and poisonous gas release that could be prevented in the oil and gas industry.

The case study part would help us to learn some big lessons from incidents and introduce adequate process safety management alongside occupational health and safety and environmental needs and integrate them into the oil and gas sectors typically of Upstream and Downstream productions.

The Final take away benefits of this PSM book consists of:

- **The Case Study**
- **The Environmental Impacts**
- **Lessons Learned from Incident**
- **Tasks Assessment (Multiple Choice Questions)**
- **Tasks Assessment (Theoretical Questions)**

ELEMENT TWENTY-SIX

THE CASE STUDY

Process Safety Leadership

Before the 20th April 2010 explosion event of BP's Deep-water Horizon Offshore drilling facility of Macondo prospects, there was a twin-hydrocarbon explosions namely, the Texas City Refinery in the United States on March 23rd, 2005, and Buncefield accident in the United Kingdom on the 11th December 2005, prompted the establishment of the **"Process Safety Leadership Group (PSLG) in 2007".**

The fundamental aim is working with the regulatory bodies to developed a functional framcwork for adequate management of leadership actions required to serve as formal guidelines for the oil and gas industry and the core function of PSLG was dealing with the hazard and risks arising from the operational activities of hydrocarbon production processes.

The Idea to include Process Safety Leadership (PSL) into the Health, Safety & Environment or HSE Management System originated from the outcome of investigations conducted on several incidents in the process industry which identifies **Incompetence (Human Errors)** as the culprit.

Knowing the origins of such catastrophes can be prevented since they are related to inadequate leadership, poor organizational safcty culture or (negative safety culture) then, training managers on status of process safety risk management within their companies would suffice.

The likelihood and severity associated with the release of petroleum hazards, containment and explosions warranted priorities should be given to the HSE Management System to evaluate the possible elimination of leadership oversight and their inadequate involvement in the process safety management in the oil and gas industry.

The solutions proposed include the need for leadership competence and a high level of engagement in the process safety management as well as establishing positive safety culture at all levels in their undertakings.

The UK's Process Safety Leadership Group (PSLG) Report states that;

At least one (1) board member should have detailed knowledge in the process safety management to enable them to provide the board with the needed advice concerning the status of the process safety risk management within the company.

Must conform with the process safety significance of the board's decisions, own to leaders' awareness, utilizing this framework for leadership and commitment in the HSE management system that requires;

- **Leadership involvement:** Leaders are required to involve with the awareness of the potential hazard and risks posed by their process industrial activities and the potential effects that may come from the board's decisions to do or not do things needed to eliminate or control the anticipated hazards and risk.
- **Competent and engaged actively:** Leaders have to be competent,

actively engaged and acknowledgeable in major accident potentials of their process entity and the needs to provide control measures together with other business risks.

The impact of these major incidents in the process industry ranges from loss of life of workers, damage to assets and property, (equipment and machinery), damage to reputation, legal consequences, and the effects on wildlife animals and marine lives alike.

• **Knowledge of plant and equipment used in their installations activities:** Leadership teams in the process industry need to be aware of different hazards and potential impacts of plant on their site which must be directly related to the life-cycle of these (systems or machinery).

Starting from the manufacturer's warehouse during acquisition, to installation, until decommissioning, analysing them for accidents they could cause is vital, major accidents do not occur by chance, they originate from process operations as they get older in use, they pose different threat with varying impacts that comes with them.

• **Ensure compliance with requirements:** This part requires directors and top managers to play major roles in promoting process safety within their organizations by assuming leadership responsibilities.

Assigning priorities to competent employees, establishing the health and safety tone for the company, while ensuring all legal responsibilities has been complied with would serves additional values.

• **Ensure the sustainability of safety culture:** Actions depicted by the process industrial leaders are noted by workers and their regular visibility becomes an integral part of the safety culture of the organization.

Employees anticipating the appearance of the top management of the company wearing Personal Protective Equipment (PPE) like helmet, gloves, mask, and others would be reinforced to abide by personal safety regulations set by the organization.

In this case, a high leadership commitment such as regular visits mixed with personally led initiatives as part of leaders' visibility would be challenging to help attain the desired positive health and safety culture.

Inadequate Leadership (Case Study)

The general knowledge of the inherent nature of the process industry, typically in hydrocarbons or petrochemicals, the awareness of working safely is paramount if safety measures required are adhered to and at all levels.

This is because most of the conclusions from investigations conducted into past incidents turned to identify preventable root causes and such when prevented, those major incidents may have never happened.

Below case studies involved past explosion in the oil and gas industry.

The Case study: DEEPWATER HORIZON INCIDENT

British Petroleum (BP)

British Petroleum (BP) is a multinational corporation headquartered in London, United Kingdom and the enterprise holds a top spot position as one of the world largest energy companies specializes in upstream and downstream production in the oil and gas industry.

BP operates globally and has a significant presence in the various aspects of the oil and gas sectors, including exploration, production, refining and distribution, other operational activities of BP also include the specialization in **Alternative Energy Sources** in the global market.

By estimation, British Petroleum is the 4th largest energy company in the globe, operating in about eighty **(80)** different countries with an average daily production ranging between **3.8 million** to **4 million** barrels of crude oil, while servicing more than **22,000** fuel stations worldwide.

The Macondo Prospect 2010

The incident of the Macondo prospect offshore oil exploration site was located at the Mississippi Canyon Block 252 of the Gulf of Mexico.

In the year 2010 April 20th, there was an explosion referred to as "Deep-water Horizon" an offshore drilling rig contracted by British Petroleum, near south Louisiana in the United States of America.

The rig blowout occurred as a result of an uncontrolled release of oil and gas from the well leading to a disaster that lasted for about 87 days.

The blowout ended up causing series of explosions which eventually led to the sinking of the Deep-Water Horizon rig, resulting the loss of 11 crew members onboard, while several others sustained multiple degrees of injuries.

The Macondo well explosion release oil into the Gulf of Mexico continuously for about 87 days before it was capped successfully.

This unfortunate incident gained international attention and was considered one of the largest environmental disasters in history with the impact of millions of barrels of crude oil being discharged into the Gulf of Mexico.

The magnitude of this catastrophic incident was massive affecting marine lives, domestic economies dependent on fishing and tourism, as well as the impact on the coastal ecosystems.

The Transportation of Deep-Water Horizon Oil Platform to the Offshore Site With an estimated value of about $560 million

During Installation of the Deep-Water Horizon Oil Platform

The Platform at Night after Installation

The Deep-Water Horizon Oil Platform ready for drilling, Macondo Prospect

A picture of the Deep-Water Horizon accident scene

Some details according to the Time Line

Deep water Horizon Accident, reported on the British Petroleum popularly known as BP by Wall Street Journal: -

At about 11 A. M on April 20th, 2010 the BP (oil and gas company leased the oil rig) (Trans- Ocean Oil Rig Operator) workers engaged in a disagreement about how to commence Shut down the (Oil Well).

According to the article from Wall Street Journal, a heavy drilling fluid has been removed from the pipes in the well, a basic sign of alarm signifies something is wrong.

Therefore, removing heavy drilling fluid before temporarily sealing up an oil well and abandoning it is often normal to the operators in the industry.

A major question arising is, whether the crew began the process without factoring into consideration all the precautions needed against the possibility of gas rising into the pipe.

Barely **11** hours after the meeting, the oil rig had a blowout, immediate cause of an uncontrolled release of oil and gas, killing **11** employees and the platform engulfed in flames, and due to offshore conditions, the oil platform sanked.

On the 20th April 2010, an explosion and fire on Trans-ocean limited drilling rig, Deep-water Horizon licensed to BP that took away 11 lives of workers also saw the drilling-platform descend beneath with an estimation of about **13,000** feet below the seabed.

Five days into the accident, April 25th an attempt to activate the rig's blowout preventer yielded no positive results.

Another effort was made on the 7th May 2010 to place a containment dome around the spewing well prove abortive when the device became ineffective for the intended purpose and such effortless resulting from the frozen hydrocarbons that clogged it.

On the 29th of May, BP sent the complex **"top kill"** manoeuvre to plug the well has failed, likewise June 2nd however, another trying as in (capping strategy) by BP yielded fruitless results.

7th July, the leakage of Oil and Natural Gas from the failed Blow out Preventer continue but some of the oil and natural gas were **"captured"** and those that **"flared"** day by day were enormous.

On the 12th of July 2010, NBC reported that up to **60,000** barrels of oil escape into the Gulf of Mexico on daily basis.

After the explosion; Now look at the Deep-Water Horizon Oil Platform; Remember the loss of lives and the financial cost

An image of blown out platform, disintegrated and about to sink to the bottom of the ocean

ELEMENT TWENTY-SEVEN

THE ENVIRONMENTAL IMPACTS

Environmental and Social Challenges

The consequences of the Deep-water Horizon oil spill had a severe impact on the environment, causing extensive damage to marine lives and local economies including coastal habitats.

The magnitude of the Macondo well blowout gave rise to a massive cleanup efforts and strategies including the deployment of booms, skimmers and dispersants aiming to contain and mitigate the wide spread of crude oil into the Gulf.

Due to the severe consequences of the Macondo's oil spill incident, awareness was raised about the potential long-term effects of such disasters on ecosystems and emphasis were made towards the needed response strategies for effectively remedy such occurrences.

The costs of marine lives

Oil-covered Pelican, from the front

Oil-covered Pelican, from the back

THE EFFECTS ON ANIMALS' LIVES IN THE WILD (Marine life)

The United States National Oceanic and Atmospheric Administration, (NOAA) Office of Marine and Aviation Operations (OMAO) operates a wide range of specialized ships and aircraft to complete the NOAA's **Environmental** and **Scientific** missions reported of a dead 25-foot Sperm Whale, turtles and other marine lives were discovered at 150 miles part of Pascagoula in Jackson County, Mississippi, 77 miles from the site of the oil spillage. See the images below.

Deep water Horizon oil spill environmental impact, Credit NOAA

This particular Sperm Whale is also known as Cachalot remains the largest kind of toothed whale and is sometimes referred to as the largest toothed predator in the Seas and larger oceans.

Despite the huge size of the body, with teeth and stronger than Dolphins and other fisheries, it couldn't survive the burning furnace and toxicity of the larger number of hydrocarbons escaping into the water, due to the explosion. Now, what about other animals?

Effect on a tooth Sperm Whale

Effect on Dolphins

Between 30th April, to 17th June 2010, other wildlife confirmed stranded by the impact of the oil spill includes **469** Turtles, and out of such a number, the greater part of about **350+** were found dead, only **37** considered to be stranded alive.

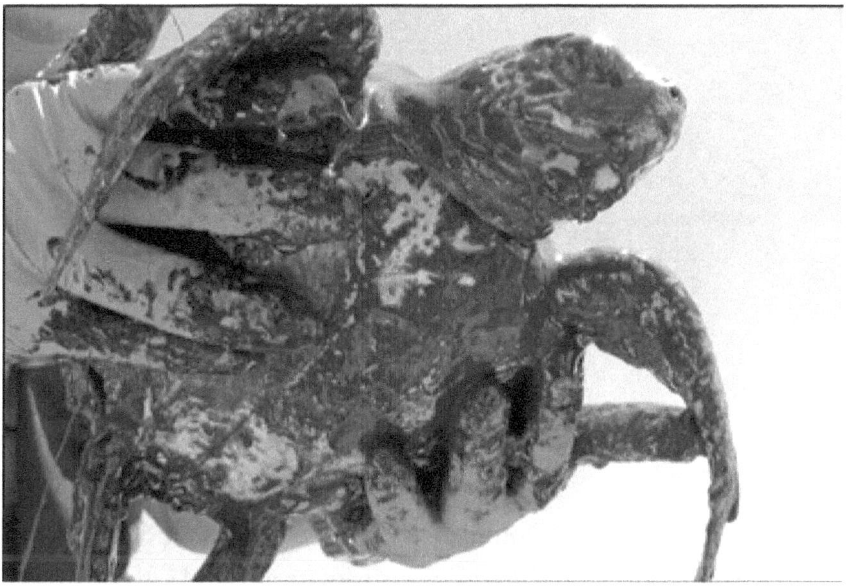

Oil-spill effect on a turtle, Image 1

Oil-spill effect on a turtle, Image 2

The struggling of Marine life in oily water

Other fisheries like Dolphins were affected too, and the report says, the total number of **46** dolphins was stranded, **43** among them ended up dead, while only 3 survived.

Some research conducted by the **Centre for Biological Diversity a non-profit membership organization in the United States,** known prominently for protecting the endangered species.

Using the prescribed legal actions, scientific-based petitions, creative media, and grassroots activism, reported about the harm caused by the oil spill, with approximations that, the following are likely to be:

1. Killed about **82,000** birds of **102** different species,
2. Harmed about **6,125** turtles, and
3. Killed or harmed about **25,900** marine animals such as Bottle nose dolphins and Spinner dolphins.

THE CLEANUP EXERCISE

After the incident of this huge explosion, the BP Company decided to take their international social corporate responsibilities further as the means of correctional activities to mitigate the harm caused by the oil spillage.

Due to that, an amount of about five (5) million United States Dollars would be allocated to the National and Wildlife Foundation, as well as revenue generated from the recovered oil spill by the corporation.

The clean-up processes initiated by the crews assigned, saw dumping of more than **1.7** million gallons of chemical dispersants into the Gulf of Mexico to reduce the toxicity of the hydrocarbons in the water.

The idea for pouring dispersants is that they are chemical substances with the ability to offer some level of support to Natural Oil-eating bacteria in the clean-up activities caused by one of the largest marine oil spills in the history of the oil and gas industry.

The substance's ability to break down the oil into droplets, however, would facilitate the microbes to consume more easily.

ELEMENT TWENTY-EIGHT

CAUSES AND LESSONS LEARNED FROM INCIDENT

The incident of Deep-water Horizon oil spillage at the Macondo prospect has significantly influenced the oil and gas industry, resulting to an increased emphasis on Safety, Regulatory Reforms, and Advancements in technology to prevent similar incident from occurring.

In addition, efforts were made to adequately improving responses to similar incident in case they happen, matters arising from the incident has remains a defining moment in the oil and gas industry's commitment to responsible and sustainable practices.

Causes and Lessons Learned

The Macondo prospect offshore drilling rig explosion was attributed to a failure in the Blowout Preventer (BOP) which is a Critical Safety Element, a device constructed and designed specifically to prevent uncontrolled releases of hydrocarbons during drilling operations or activities.

However, investigations conducted into the incident by authorities revealed other multiple contributing factors such as inadequate well design, interpretation failures (failure to pass a good judgement on pressure test results) and inappropriate cementing of the wellbore.

The Deep-water Horizon explosion also prompted a re-evaluation of Safety Practices, Emergency Response Capabilities and Strategies, as well as Regulatory Oversight within the oil and gas industry and huge losses originated from the oil spill ranges from Legal Actions against BP, Substantial fines and Settlements was enormous.

Additionally, the blowout also propelled series of changes in the oil and gas sector covering regulations and standards alongside the enhancement in **Safety** and **Environmental Protection** from such disasters.

Lessons Learned from the impact of numerous accidents especially, of (Hydrocarbon related explosions) have put Competent Authorities, Regulatory Bodies, and Institutional units to developed safety improvements.

Other participants include Industrial Associations, and corporate entities, with their leaderships and jointly devised series of structures needed as the **fundamental framework function** for the HSE management systems in the process industry (PI).

The development and implementation of the HSE management systems then became an integral part of the Oil and Gas Industry and has since been rooted with a wider spread anticipation across regions globally.

The Fundamental Framework Function now accounted for managing all **major hazards**, and any other hazards using **systematic approaches designed** to enable leadership's ability to **further improve health** and **safety performance** in the industry.

The HSE management system serves as an essential mechanism to facilitate companies' operational activities in the process industry to use the best practice of the HSE standards, with prospects of;

- Maximizing the protection of employees, and other people from injury and hazards arising from work,

- Minimizing the potential legal actions against corporations,
- Adequate systems for managing the use or production of "Highly Hazardous Chemicals" for example, **hydrocarbon groups**,
- Ensuring Process Safety Responsibilities at all times,
- Utilizing Management of Change (MOC),
- Obtaining Process Safety Management Information from the right sources, and
- The required consistency in integrating technologies, methodologies, and management practices timely by using the interrelated activities in the HSE management systems to determine the level of interdependence among them based on the objectives and the application of good design principles.

Major Incidents References: Benefits of Learning from Incidents

Bhopal, India, **(1984/1985).** The release of Toxic gas that killed about **2,700** people, nearly **50,000** were the seriously affected, while the less affected individuals are considered to be around **1,000,000** by approximation.

Piper Alpha, **(1988)**. The explosion resulted in killing about **165** men onboard Piper Alpha scenario in Scotland's North East part **120** miles from Aberdeen. An estimated largest oil rig blown out in the oil and gas industry in terms of death toll.

Texas City, United States **(2005).** The Texas City refinery explosion where flammable hydrocarbon vapour cloud ignited and violently exploded, killing about **15** people (employees) and injured several others around **180** individuals and causes severe property damage to the refinery.

Buncefield, UK **(2005)**. This incident resulted about **40** injuries, with a widespread damage.

Deep water Horizon Oil Spill, **(2010)**. This blown out of BP Macondo prospect also known as the Major oil spill in the Gulf of Mexico, claimed **11** human lives, property damage of huge costs, enumerated loss of marine life in the sea.

Learning from major catastrophic incidents mentioned above could help Oil and Gas companies to utilize the following:

A. The use of Permit-to-Work Systems

B. Safety Management (Safe Operating Procedures, sops)

C. Rig Design Improvements

D. Suitable Maintenance Systems

E. Conduct Adequate Safety Training

F. Perform Safety Audits and maintain Positive Safety Culture

- A good safety culture practice depicts absolute shared attitudes and beliefs based on workplace behaviour of employees, top management, safety practitioners, engineers, operators, maintenance team, and so on.

Organizations with positive safety culture can show the following observations:

ITEMS	SOME OF THE POSITIVE SAFETY CULTURE INDICATORS
1	High level of regards for health and safety (HSE) across board
2	Good knowledge about hazards and risk is a shared responsibility by all workers, subordinates and superiors alike
3	Adopting the same positive attitude at all levels and by all employees and could been seen in the board member(s) – Company's Top Officials
4	Exhibiting high level of ownership, taking responsibility and accountability for the success or failure in health, safety and environmental management could easily be identified with the company.

ELEMENT TWENTY-NINE

TASKS ASSESSMENT (MULTIPLE CHOICE QUESTIONS)

Complete the following tasks assessment (Q1 – Q50)

Q1. What does the abbreviation word MOC stands for?

a. Management of Chance

b. Management of Company

c. Management of Change

d. Management of Changes

Q2. Assign the full meaning to the acronym PSMS

a. Peoples Safety Management Sheet

b. Process Safety Management Systems

c. Peoples Safety Management Systems

d. Process Safety Management Sheet

Q3. What does the acronym ALARP mean

a. As Long As Reasonably Practicable

b. As Law As Reasonably Practicable

c. As Lone As Reasonably Practicable

d. As Low As Reasonably Practicable

Q4. Provisions for portable drinking water, suitable washrooms, first aid kits for employees' comfort are known as

a. Welfare

b. Safety

c. Health

d. Health and Safety

Q5. The act of protecting workers, contractors, suppliers and community members from occupational or work-related ill health such as mental sickness and inhalation of poisonous substance is known as

a. Health and Safety

b. Welfare

c. Safety

d. Health

Q6. When designing health and safety systems, it is highly recommended that provisions should be made to report……………………to the relevant authorities

a. Ebola Pandemic

b. Disaster Occurrences

c. Dangerous Occurrences

d. Natural Disasters

Q7. What is the full meaning of the acronym KATE?

a. Knowledge, Authority, Training and Experience

b. Knowledge, Ability, Training and Experience

c. Knowledge, Assessment, Training and Experience

d. Knowledge, About, Task and Experience

Q8. The chemical reaction between oxygen, fuel and heat is known as

a. Fire

b. Element of fire

c. Fire tetrahedron

d. Explosion

Q9. Fire processes from induction, to the ignition, to the growth, and then to the steady stage before proceeding to decay. What is the name for this sequence?

a. Wildfires

b. Chain Reaction / Tetrahedron

c. Combustion Stages

d. Fire elements combination

Q10. What is the full form of fire

a. Find, Inform, Restrict, and Extinguish

b. Oxygen, Fuel, and Heat

c. A and D

d. Find, Inform, Restrict and Evacuate

Q11. What is the full meaning of the acronym FPSO?

a. Floating, Production, Storage and Offloading

b. Floating, Processes, Storage and Offloading

c. Floating, Producing, Storage and Offloading

d. Floating Processing Storage and Offloading

Q12. Assign the full meaning to the term 'PSM' in the process industry (PI)

a. Peoples Safety Management

b. Production Safety Management

c. Process Systems Management

d. Process Safety Management

Q13. When selecting personnel for occupational health and safety representative position, it is highly recommended to consider using the requirements below, which are they?

a. Knowledge and Ability

b. Training and Experience

c. A and B

d. None of the above

Q14. What does the acronym PPE stand for?

a. Protecting People Everywhere

b. Personal Protective Equipment

c. Peoples Protection Equipment

d. Protecting Personal Equipment

Q15. The act of protecting workers from physical injury at the workplace is termed as

a. Safety

b. Health

c. Health and Safety Management

d. Near Miss

Q16. What is the most suitable name for an individual who possesses competence, leadership qualities, good observation skills, and can pass judgement on risks and hazard

a. Health and Safety Representative

b. Safety Supervisor

c. Safety Officer

d. Safety Engineer

Q17. Personnel holding safety supervisor position entails the role of being responsible for supervision and leading a team to work safely

a. Maybe

b. False

c. True

d. I'm not sure

Q18. A set of rules and regulations meant to regulates behavior in the society is referred to as

a. Criminal Law

b. Civil Law

c. The Law

d. Vicarious Liability

Q19. Which type of law require passing judgement should highly base on the balance of profitability

a. Criminal Law

b. Civil Law

c. The Law

d. Vicarious Liability

Q20. Which law allows employers to be liable for an offense, even when it was committed by a worker(s)

a. Criminal Law

b. Civil Law

c. The Law

d. Vicarious Liability

Q21. In occupational safety and health practices, improvement and prohibition notice are enforcement under

a. Methods of compliance

b. Balance of profitability

c. Beyond reasonable doubt

d. The HSE Law

Q22. In the oil and gas industry, the process of drilling, exploration and production processes of crude oil and natural gas is known as

a. Downstream

b. Upstream

c. Midstream

d. None of the above

Q23. The operational activities of processing, refining and distribution of crude oil and natural gas products is termed as

a. Hydrocarbons processes

b. Midstream

c. Downstream

d. Upstream

Q24. The transportation, storage and logistics section that serves as an intermediary between upstream and downstream production stages of oil and gas is known as

a. Hydrocarbons processes

b. Downstream

c. Midstream

d. Upstream

Q25. The upstream, downstream and midstream are the oil and gas industrial terms used to describe different stages of production

a. False

b. True

c. None of the above

d. All of the above

Q26. Utilizing the act of secondary containment, implementing security measures for access control, and systems to monitor leaks detection are procedures for

a. Hydrocarbons production

b. Crude oil and natural gas production

c. Design and construction of storage facilities

d. Hydrocarbons safe storage management

Q27. Conducting regular maintenance and inspection, design and construction of storage facilities, measures for fire protection and protection of groundwater are measures used in the oil and gas industry for

a. Hydrocarbons safe storage management

b. Crude oil and natural gas production

c. Upstream, downstream and midstream sectors

d. Design and construction of storage facilities

Q28. Select the most appropriate answer to describe the two types of boilers that are commonly used in the oil and gas industry

a. Fire-tube boiler

b. Steam generation

c. Water-tube boiler

d. A and C

Q29. Steam management are made of 3 basic components, choose the most suitable answer that justify this statement

a. Temperature control

b. Steam generation

c. A, B and D

d. Pressure control

Q30. Steam distribution consists of two fundamental components, which are they?

a. Pressure relief and valves

b. C and D

c. Steam traps

d. Piping systems

Q31. Class C fires are best described as fires involving

a. Flammable gases

b. Butane

c. Propane

d. None of the above

Q32. Fires are classified as category A when it involves

a. Organic solid materials

b. Wood and paper

c. Fabrics and plastics

d. All of the above

Q33. Class D fires are made up of all of the following except

a. Aluminium

b. Lithium and magnesium

c. None of the above

d. All of the above

Q34. Flammable liquids are classified under which category of fires

a. Class A

b. Class B

c. Class C

d. Class D

Q35. What is the most appropriate fire extinguishing agent for combustible metals

a. Lithium and magnesium

b. Aluminium and sodium

c. Specialized Dry Chemical Agent Designed for Metallic Fires

d. None of the above

36. What is the extinguishing agent for class A fires

a. Foam

b. Water-based solutions or Water

c. Carbon dioxide

d. Dry chemical powder (DCP)

Q37. Select the most suitable extinguishing agent for class B fires

a. Dry chemical agents

b. Foam

c. Carbon dioxide

d. All of the above

Q38. The systematic approach for efficient and effectively control hazards is known as

a. Hierarchy of controls

b. Elimination of hazards

c. Engineering controls

d. Technical controls

Q39. Which of the following best describes the accurate implementation of the hierarchy of controls

a. Control hierarchy

b. Elimination, Engineering, Procedural controls, and PPE

c. Technical controls

d. Administrative controls

Q40. Which of the following statements justifies the accurate arrangements of control hierarchy

a. Hierarchy of controls

b. PPE

c. Elimination, Technical, Administrative and Personal Protective Equipment

d. Procedural and Engineering controls

Q41. What is the flash point of an extremely flammable substance

a. Below 0°C

b. 0° - 21°C

c. 22° - 55°C

d. Flammable range

Q42. Assign flammable category to substance that are flammable

a. Flammable range

b. 22° - 55°C

c. O° - 21°C

d. Below 0°C

Q43. According to the limits of flammability illustration, 100% AIR and 0% FUEL represent

a. Very Rich

b. Flammable Range

c. Very Lean / Too Lean

d. Fuel, Gas, Air and Vapour Mixture

Q44. Choose the most suitable category of flammability for highly flammable substance

a. 0° - 21°C

b. Flammable categories

c. Below 0°C

d. 22° - 55° C

Q45. Studying the diagram of flammability limits, 0% AIR and 100% FUEL means

a. Flammable Range

b. Very Lean

c. Fuel, Gas, Air and Vapour Mixture

d. Very Rich / Too Rich

46. Learning from the risk estimation and risk ranking chart box, what does the risk assessment in the red colour zone represent

a. 1 - 4 Low risk

b. 5 - 10 Medium risk

c. 12 - 25 High risk

d. None of the above

47. Assign risk assessment interpretation to the figures in green colour zone

a. 1 - 4 High risk

b. 1 - 4 Low risk

c. 12 - 25 High risk

d. 5 - 10 Medium risk

Q48. What are the technical terms used in safety when determining the level of risk

a. Likelihood

b. Severity

c. Likelihood and Severity

d. Risk ranking and risk estimation

Q49. According to the 5 by 5 metrics used for risk estimation and risk ranking, select the most appropriate representation for the figures in the yellow colour zone

a. 5 - 10 very unlikely

b. 5 - 10 Medium risk

c. 5 - 10 Unlikely / May occur

d. None of the above

Q50. The most appropriate approach for developing safe systems of work (SSOW) is to consider

a. People, Equipment, Materials and Environment

b. Identify hazards and evaluating the risks

c. Define safe methods, implementing the SSOW, and Conduct review

d. All of the above

ELEMENT THIRTY

TASKS ASSESSMENT (THEORETICAL QUESTIONS)

Complete the following tasks assessment (Q1 – Q30)

Q1. Defined the following terms in occupational health and safety practices

i. Health

ii. Welfare

iii. Safety

iv. Dangerous Occurrences

v. Class 'D' fires

Q2. Describe the following in health and safety terms

i. Accident

ii. Near Miss

iii. Incident

iv. Accident Ratio

v. Fire-tube boiler

Q3. Explain the following items in their abbreviation form into full meaning in brief. For example, PJSC full meaning in brief as a Public Joint Stock Company.

i. MOC

ii. PSM

iii. ERP

iv. HSE

v. Water-tube boiler

Q4. Write down the full meaning of the following acronyms

i. FPSO

ii. ALARP

iii. SSOW

iv. PSMS

v. POPMAR

Q5. Write down three (3) advantages and two (2) disadvantages of an FPSO

Q6. Describe the following recommendations by UK's Process Safety Leadership Group

i. Leadership involvement

ii. Competent and engage actively

iii. Knowledge of plant and equipment

iv. Ensure compliance with requirements

v. Ensure the sustainability of safety culture

Q7. Explain Process Safety Management in details?

Q8. Explain Management of Change in details?

Q9. State and explain any 5 components of Floating, Production, Storage and Offloading (FPSO) vessel.

Q10. Define the following terms under health, safety and environment law

i. Criminal law

ii. Vicarious liability

iii. Civil law

iv. Improvement notice

v. what is law?

Q11. Describe the following

i. Prohibition notice

ii. Offshore

iii. "Replacement In-kind"

iv. Onshore

v. Key activities of upstream

Q12. Write down the full meaning of the items below

i. HAZOP

ii. HAZID

iii. FMEA

iv. OSHA

v. ESD

Q13. Briefly describe the following items

i. Establish overall objectives

ii. Set targets

iii. Monitor progress of indicators

iv. Frequent review

v. Class 'C' fires

Q14. State and explain any five (5) steps suitably for oil and gas companies when designing Emergency Response Procedures

Q15. Explain the 4 steps that can be used to conduct basic investigations

i. List the four methods of basic investigations in an orderly manner

Q16. Write down four (5) recommendations given by OSHA as an ideal step-by-step to follow when implementing MOC by process entity.

Q17. State and explain any five (5) steps to store hydrocarbons safely.

Q18. Explain the following

i. Upstream

ii. Midstream

iii. Asset Integrity

iv. Downstream

v. Class 'B' fires

Q19. Write down the full meaning of the following HSE regulatory bodies and provide at least 3 of their parent organizations alongside their dates of inception.

i. HSE UK

ii. WSA

iii. ILO

iv. OSHA

v. Homicide European

Q20. Learning from duo-index, you are required to assign an index and risk levels of the following grade according to Dow Index Risk Classification Table.

Grade

A –

B

C –

D –

E –

Q21. Describe the following

i. Process flow diagrams

ii. Equipment specifications

iii. Cause and effect diagrams

iv. List of controls

v. Class 'A' Fires

Q22. State and explain 5 start-up techniques

Q23. Describe 5 shut-down methods

Q24. State and explain five (5) commissioning procedures

Q25. Define five (5) decommissioning processes

Q26. Explain any five characteristic steps for a standardize permit-to-work in general for the process entity that you worked for?

Q27. Write down any five (5) items that a permit issuing authority must be familiar with

Q28. Explain any 5 roles and responsibilities of an individual assigned as performing authority of PTW in an oil and gas installation

Q29. Write down five key elements of safe systems of work

Q30. What is fire?

A. Explain the following

i. Fuel

ii. Heat or Light

iii. Oxygen

iv. Fire triangle

www.ingramcontent.com/pod-product-compliance
Lightning Source LLC
Chambersburg PA
CBHW031520180726
48282CB00056B/162
9789360064372